Getting the most from your GPST years

SUCCESS IN MEDICINE SERIES

Getting the most from your GPST years

The truth about hospital posts, the ePortfolio, working as a registrar, the dreaded exams, plus finding and getting a job at the end of it all...

WRITTEN BY

Dr Lucy Blunt, MRCGP

SENIOR MEDICAL ADVISORS

Dr Katharine King
Dr Mike Houghton
Dr Kay Bridgeman

OXFORD
UNIVERSITY PRESS

OXFORD
UNIVERSITY PRESS

Great Clarendon Street, Oxford OX2 6DP
United Kingdom

Oxford University Press is a department of the University of Oxford.
It furthers the University's objective of excellence in research, scholarship, and education by publishing worldwide. Oxford is a registered trade mark of Oxford University Press in the UK and in certain other countries

Published in 2011
MDU edition published in 2012

Impression: 1

British Library Cataloguing in Publication Data

Data available

Library of Congress Cataloging-in-Publication Data

Data available

ISBN 978–0–19–966253–1

Printed in Great Britain
on acid-free paper by
Printed and bound by CPI Group (UK) Ltd, Croydon, CR0 4YY

This is for Patrick,
who during the writing of this book
grew to become a beautiful butterfly,
from just a tiny little caterpillar.

Foreword

The moment you embark on your GP training is an exhilarating time. Having met new challenges as a junior doctor on the wards for the first time, by this stage you will have a growing confidence in your core clinical skills and be ready for the next challenge.

As a GP trainee, the next three years will be an opportunity to apply your clinical skills in a fascinating and rewarding field, working with experts and learning all the time.

The current system of GP training provides a structured career path, although there is always pressure to stand out from the crowd with each assessment, examination and placement so that you have the best chance of getting your dream job at the end of your training years.

Reading the practical advice in this book will leave you in no doubt that you need to think ahead in order to succeed. For example, it's better to commit time to regularly documenting your progress and achievements in your ePortfolio to impress the ARCP panel, rather than attempting any late-night cramming.

Remember that achieving your potential will also depend on how effectively you prepare for the various assessments you will face. At every stage in your training you will need to demonstrate you have the professional and personal skills required to be a GP so it's important to obtain as much feedback as you can about your consultation skills during your placements and reflect on what you can improve. You can find valuable advice within these pages on how to approach the role play scenarios you might face during the clinical skills assessment in the MRCGP examination.

It's only natural to have practical concerns about other aspects of your life during your training so this book is not solely about the pursuit of a career in general practice. There are also welcome tips on topics such as controlling your finances and achieving that tricky work-life balance.

Whether it's your ambition to join a GP practice straight away or you are considering other options, I'm confident that this book will see you through your training years. I am also delighted that as an MDU member you will have the chance to benefit from the exclusive additional content in this bespoke edition, including practice multiple choice questions for the Applied Knowledge Test and top ten tips for training.

This is just another example of the MDU's commitment to supporting members throughout their careers. Our expertise means we are able to help you with your continuing professional development; keep you informed about the latest guidance on good practice in our journals and on our website; and provide individual ethical and legal advice through our 24-hour freephone helpline. As a medical defence organisation, we aren't only there for our members in times of trouble; we are also there to provide a valued and trusted source of support, advice and assistance during your professional life.

Thank you for choosing the MDU and good luck with the next chapter in your career.

Dr Christine Tomkins
BSc (Hons) MBChB (Hons) DO FRCS FRCOphth MBA FFFLM FRCP
Chief Executive
The Medical Defence Union

Preface

A little about why I wrote this book

I joined the Coventry and Warwickshire VTS Scheme in 2006 after completing my house officer year and one year as a medical SHO at Warwick Hospital.

Over the last three years of my VTS training I have been fortunate enough to have received a great deal of helpful advice from previous trainees and trainers, however the more valuable lessons have been learnt through the harder ways of 'experimentation', 'trial and error', and 'just getting on with it'.

When our VTS cohort commenced the training scheme in 2006, the ePortfolio concept was just a twinkle in the RCGP's eye and it wasn't until late 2007 that we were suddenly faced with the daunting task of trying to get consultants to complete online assessment of our DOPS and mini-CEX without them sniggering. Of course, those that had done F2 placements were all too used to this, but for those of us who were still very much in the old school routine of being house officers and SHOs this was a very scary idea.

Luckily since then the ePortfolio has become much more sophisticated and far easier to navigate. As with all new things, over time, both trainees and trainers have not only learnt how to use it to their advantage but have also succeeded in keeping up with its ever changing profile and new ideas.

But it wasn't just the ePortfolio which was new for us. In 2006 the MRCGP assessment process switched to that of involving not only an ePortfolio but also an AKT, a CSA, and a WBPA which would replace the old MCQ, videos, and viva. No more posing gaily at the camera... or so we thought... (later we were to find out that COTs changed all of that!) It seemed like a much more effective way of proving your ability, although at a significantly higher cost than the old assessment system (but we won't go on about that!).

Gradually, books and courses began to appear to aid us along our assessment journey and it wasn't long before the first successful cohorts started to come out the other side of the new MRCGP process.

It's definitely been a bumpy ride as a trainee over the last three years, in the midst of so much change, but we have learnt a great deal during that time. Not only have we finally learnt how to successfully complete the ePortfolio but we have also learnt to value the new assessment process and its positive effects on GP training. There are many books out there that will give pages of AKT questions or CSA scenarios for you to practise, but there is nothing that supports you through the whole process of GP training from start to finish and my aim is that this book will change all of that.

I hope that this book not only answers some of the common questions that you may encounter at various stages of your training, but that it also helps you to avoid making some of the mistakes that we made along the way and therefore leads you to successful accomplishment of the MRCGP. The content is based on a combination of fact, opinion, whinges expressed over coffee at VTS teaching, and some handy hints and tips that people going through it have kindly shared with me on social nights out.

So scribble in it, spill tea on it, and make it your own. This is your book to accompany you through the next three to five years of your career and I hope that it brings you a couple of laughs along the way.

GP training is a fantastic opportunity that should be embraced with both arms and given the credit that it deserves for producing such well-rounded competent doctors.

Enjoy!

Fees and other costs are accurate as of 2010, but may since have changed.

Dr Lucy Blunt
GPwSI Gynaecology
MRCGP LoCMED LoCIUT LoCSDI
DFSRH DRCOG MBChB

Acknowledgements

This book would never have been possible without the encouragement and guidance of my incredibly supportive medical advisors and Training Programme Directors; Katharine, Mike, and Kay, whose experience and wisdom has been invaluable to me over the last three years. I am so grateful for their speedy peer reviewing and constructive criticism which has hopefully made this book the best that it could possibly be.

My thanks must also go to Fiona and Katy at OUP, for not only giving me the opportunity to publish this book but for all of the helpful phone calls and emails along the way.

I am also extremely grateful to the Coventry and Warwickshire GP trainees who have contributed to this book (names listed below) by way of providing quotes, writing about their experiences, and giving me inside information on their training years. In particular I would like to thank the 2009 ST3 crew who believed that I could do it and allowed me to pick their brains over sandwiches and cake at teaching.

To the other two musketeers, Alice and Hannah, with whom I have laughed, cried, and whinged over endless cups of tea during the last three years, in an attempt to get through our training without losing the plot too much. They have been my true soul mates and have provided that extra bit of encouragement and energy when things got tough.

To my ST1 trainer Cristina, for reminding me to 'never cut corners' and for inspiring me to put my heart and soul into general practice from day one—I owe you an even bigger bunch of flowers than before.

To Sarah, for her insight into becoming a parent and her brilliant contribution to Chapter 16 (not to mention the goodies which she sent that got me through the early weeks after Patrick was born).

And of course I must thank my fantastic parents, without whom I would never have had the opportunity to study medicine and become a GP. Thank you to my family for your continued support over the last 10 years of ups and downs and for the inside info on the world of publishing (thanks Aimee!).

To my gorgeous little boy Patrick, for entertaining himself under the baby gym and allowing me to write this during the short times which he napped. I hope that he will one day read this book and be proud that he was such a big part of what I have achieved.

And finally, last but definitely not least, to my long suffering husband Pete who didn't instantly request divorce when at three months pregnant I announced that now was my time to prove the theory that 'everyone has a book in them'. I couldn't have done it without you, xx.

The contributing Coventry and Warwickshire VTS Trainees

Dr Becca Hadzikadunic, Dr Peter Reeves, Dr Samir Khan, Dr Hannah Packman, Dr Alice Bird, Dr Sarah Hall, Dr Debjit Thakurta, Dr Tesan Hadzikadunic, Dr Julie Harper, Dr Natalie Wookey, Dr Melissa Robertson, Dr Rachel Barnes, Dr Sonia Chery, Dr Hannah Collier, Dr Liz McEvoy, Dr Farheen Naqri, Dr Michelle Rogers, Dr Laura Bennett, Dr Omar Mohammed, Dr Souad Hamdy, Dr Fauzia Zafar, Dr Cat Middlemiss, Dr Laura Stirling, and Dr Emma Cope.

And other contributing GPs

Dr Sarah Forbes, Dr Cristina Ramos, Dr Jonathan Leach, and Dr Dave Rapley

Contents

How to use this book

I have tried to write this book in a vaguely chronological order, however, depending on at which points in your training you decide to do what, this may not work for you. Instead feel free to pick and choose which chapters you read and which you leave out. I would suggest that if you've already got a place on a training rotation then perhaps skip Chapters 1 and 2 as they will probably not be very useful to you.

I'd really like you to make this book your own by annotating it, highlighting important sections, and folding over pages that are particularly helpful, in attempt to make the next three years run as smoothly as possible. I have also provided a table for you to keep a log of all of your user names, membership numbers, and passwords so that you don't have to constantly keep clicking on the 'Forgotten your password?' link for every website you need to access (like I did!). There is also a table dedicated to mapping out your training so that you can plan in advance for the years ahead.

Whilst this book is by no means comprehensive in terms of telling you everything you need to know (after all, spoon-feeding gets you nowhere), I hope that it answers some of the more common queries that you may have during your training. The factual information within this book was correct as of July 2010; however, given how much GP training has adapted over the last few years, there are bound to be some things which will change and therefore it is essential that you regularly check the RCGP website for new information.

My important details

GMC number
NTN number
MPS/MDU number
BMA membership number
BMA login
BMA password
ePortfolio username
ePortfolio password
RCGP member's login
RCGP member's password

Planning your training (if a full-time trainee)

Use the following table to plan out what you hope to achieve over the next three years. Think about things such as AKT, CSA, courses, OOH, diplomas, and educational supervisor reviews. Don't forget to bear in mind any personal or social events that may influence your decisions.

	ST1	ST2	ST3
August			
September			? CSA
October			? AKT
November	1st ES review	3rd ES review	5th ES review?CSA
December			
January			? AKT
February			? CSA
March			
April			
May	2nd ES review	4th ES review? AKT	Final ES review AKT/CSA last sitting
June			
July			

Gobbledegook—some of the commonly used acronyms and what they all mean

If there's an acronym that you can't find here then search the glossary of 'GP training and assessment terms' found on the RCGP website[1].

AED	Automated External Defibrillator
AiT	Associate in Training
AKT	Applied Knowledge Test
APMS	Alternative Provider Medical Services
ARCP	Annual Review of Competence Progression
A&E	Accident and Emergency
BJGP	*British Journal of General Practice*
BMA	British Medical Association
BMJ	*British Medical Journal*
BNF	*British National Formulary*
CMT	Core Medical Training
CbD	Case-based Discussion
CCT	Certificate of Completion of Training
COGPED	Committee of GP Education Directors
COT	Consultation Observation Tool
CPD	Continued Professional Development
CRB	Criminal Records Bureau
CPR	Cardiopulmonary Resuscitation
CS	Clinical Supervisor
CSA	Clinical Skills Assessment
CV	Curriculum Vitae
DDRB	Doctors and Dentists Review Body
DES	Directly Enhanced Service
DFFP	Diploma of the Faculty of Family Planning (now replaced by DFSRH)

DFSRH	Diploma of the Faculty of Sexual and Reproductive Health
DOH	Department of Health
DOPS	Direct Observation of Procedural Skills
DRCOG	Diploma of the Royal College of Obstetrics and Gynaecology
DWP	Department of Works and Pensions
EDD	Estimated Date of Delivery
ENT	Ear, Nose, and Throat
ES	Educational Supervisor
EWTD	European Working Time Directive
FY1	Foundation Year 1
FY2	Foundation Year 2
GMC	General Medical Council
GMS	General Medical Services
GPC	General Practice Committee
GPST	General Practice Speciality Trainee
GPwSI	GP with Special Interest
GWC	General Whitley Council
JCPTGP	Joint Committee for Postgraduate Training in General Practice
LARC	Long-Acting Reversible Contraception
LES	Locally Enhanced Service
LMC	Local Medical Committee
LOC	Letter of Competence
LTFTT	Less Than Full-Time Trainee
MA	Maternity Allowance
MCQ	Multiple Choice Questions
MDU	Medical Defence Union
Mini-CEX	Mini Clinical Evaluation Exercise
MSF	Multi Source Feedback
MPS	Medical Protection Society
MRCGP	Member of the Royal College of General Practitioners
NIC	National Insurance Contribution
NICE	National Institute of Health and Clinical Excellence
NROGP	National Recruitment Office for General Practice
NTN	National Training Number
OOH	Out Of Hours
O&G	Obstetrics and Gynaecology
PBC	Practice-Based Commissioning
PCT	Primary Care Trust

PCO	Primary Care Organization
PCME	Primary Care Medical Educator
PDA	Personal Digital Assistant
PDP	Personal Development Plan
PLAB	Professional and Linguistics Assessment Board
PMETB	Postgraduate Medical Education and Training Board
PMS	Personal Medical Services
PSQ	Patient Satisfaction Questionnaire
QOF	Quality and Outcomes Framework
RCGP	Royal College of General Practitioners
RCOG	Royal College of Obstetricians and Gynaecologists
SAC	Selection Assessment Centre
SDL	Self-Directed Learning
SHO	Senior House Officer (equivalent to FY2 onwards)
SHA	Strategic Health Authority
SMP	Statutory Maternity Pay
SPP	Statutory Paternity Pay
SSP	Statutory Sick Pay
ST1	Specialist Trainee Year 1
ST2	Specialist Trainee Year 2
ST3	Specialist Trainee Year 3
UK FPO	United Kingdom Foundation Programme Organisation
VTS	Vocational Training Scheme
WPBA	Workplace-Based Assessment

REFERENCE

1 RCGP. *Glossary of GP Training and Assessment terms*. Available at: http://www.rcgp-curriculum.org.uk/info__resources/glossary_of_terms.aspx (accessed 14 August 2010).

PART I

Pre VTS

CHAPTER I

The million dollar question: do you want to be a GP?

Introduction

So herein lies the most important chapter of all. Do you really want to be a GP? It's a question that by now you should hopefully be able to answer 'yes' to, although you wouldn't be the first person in the world to be honest and say that you 'think so but haven't really had enough experience of it yet to be sure'.

Unlike our VTS cohort you are likely to have already experienced some general practice during your foundation years, however there will still be a group of people for whom the only real snap shots they have had are the few weeks or months that they spent in GP placements whilst at medical school. I have vague memories of those days and all I seem to remember is that there seemed to a great deal of tea drinking and chatting. Unlike the disgruntled hospital consultants however, they were all very friendly, seemed to have a good work–life balance, and wore jumpers rather than suits.

The 'end of GP assessment' that we had to undertake was ridiculously scary, as you just knew that anything from a sore throat to a myocardial infarction could walk through the door and you would have to take a structured history, examine the patient, formulate a differential diagnosis, and then discuss a management plan. But of course we didn't really have to do all of that—it was just that as keen medical students wanting to do well in our placements, we automatically presumed that this level of expertise was expected. In truth, it's not until your GP membership exams that anyone really expects you to do this and actually at that stage you could pass the exam by just introducing yourself, making good eye contact, and being nice to the patient. If only they had told us that in advance!

From an early stage in medicine however, it had been drummed into us that a large proportion of medical graduates would end up working as GPs. To me, that was no big issue as I had always thought myself to be quite suited to working in the community, dealing with not just patients but their families and their social circumstances and being able to provide that continuity of care which is so unique to general practice. But to others sat in our lecture hall on that first grey September Monday morning, being given our introductory lecture by the Medical School Dean, it was a disaster to hear that over 50% of us would head down the 'generalist' route instead of winning Nobel Prizes as heart surgeons. And that was because general practice seemed to have a stigma attached to it and I'm not really sure why. I think that having to know a reasonable amount about lots of things had wrongly led people to believe that GPs don't really know much about anything, and as a result the route to becoming a specialist had always seemed the more ambitious way to go. I, however, completely disagree with this stereotype, as do the vast majority of my peers and colleagues. Being able to put your hand up and admit when you have reached the extent of your knowledge and refer onwards is an essential quality to have, second only to the all-important skill of being able to cope with uncertainty on a daily basis, but we'll talk more on that later.

Deciding whether or not general practice is something for you is a decision that only you can make and the answer you decide upon now may not necessarily be the one that you thought you'd make five years ago. All I can say is that it's great fun, no two days are ever the same, and juggling your work–life balance is perhaps easier than with other specialities. You'll rarely save lives (in the acute resuscitation sense!) so if that's where you get your kicks then it's probably not for you. If, however, you're after a career which offers variety and great job satisfaction then this might just be the one. In the words of Professor David Haslam, Royal College of General Practitioners (RCGP) President, 'your life as a GP is only limited by your imagination and your energy'[1] which just about sums up how flexible and exciting general practice can be.

Despite reading all of the above and having talked it through with your family, friends, and colleagues, there are likely to still be many unanswered questions in your mind, a few of which I will endeavour to guess and briefly answer within this chapter. Further information can be also gained from **CHAPTER 11, 'YOUR FIRST GPST POST AND SOME BASIC TIPS ON HOW TO SURVIVE'**.

About being a GP

What does an average day working as a GP involve?

Of course, when you look at it nicely spread out like this, it all looks like a piece of cake. Unfortunately, however, very few days tend to run in such a structured way and most of the time you will find yourself trying to sign prescriptions with one hand, whilst

Table 1.1 A typical day at the office

Time	Activity
7.30–8.00	Arrive at surgery. Make a cup of tea! Fire up computer and check lab results—act on any that are urgent. Eyeball surgery lists for the day
8.00–8.30	Surgery starts. Patients booked at 5–10-minute intervals for between 2 and 3 hours
11.00–11.30	Coffee break. Meet with colleagues to discuss visits and post. Finish filing lab results
11.30–2.00	Visits (may be few or endless!), dictating letters, making phone calls to either patients or hospital colleagues, signing prescriptions, practice meetings... and lunch if you're lucky!
2.00–5.00	Surgery starts. Patients booked at 5–10-minute intervals for between 2 and 3 hours (some with a break for coffee and some without!)
5.00–6.30	Urgent visits, more paperwork or phone calls, more prescriptions to be signed... and home if you're lucky!
6.30–8.00	Extended hours surgery depending on what your practice offers

talking on the phone to a consultant colleague with the other, all at the same time as using your elbow to move the mouse as you try and scroll through the patient's notes on the computer. Every surgery will vary hugely on what is and isn't a normal day and in fact if you are either on call or duty doctor for that day it may well be a very different story! The picture will also vary depending on whether you are a salaried GP, a partner, a locum, a GP trainer, or a registrar, as each of these roles will have additional things to factor into their day.

How long will it take me to become a fully qualified GP?

This all depends on which route you follow and whether or not you have any career breaks (for example, to have children, travel the world, do research, or go abroad to do some voluntary work). At present, GP training is commenced after completing the FY1 and FY2 years and comprises of three years of specialist training, of which a large proportion is spent actually working in primary care.

There is, however, talk at the College of increasing the length of the training scheme to five years, in line with the hospital specialities. The aim of this is to allow trainees to spend more time in general practice without missing out on the benefits of experiencing the hospital posts. It would also provide trainees with more time to learn the administrative and managerial aspects of general practice which at present are perhaps only picked up once you are working as a fully fledged GP. At this stage it is uncertain as to when or if the change to five years will take place, but no doubt once this information becomes available it will be widely discussed amongst trainers and trainees during VTS study days.

> 'If your long-term plan is to do general practice then get on to the speciality training programme as soon as possible. Don't waste time doing other speciality training that won't count towards general practice.'
>
> ST2

How much will I get paid?

At present, the payment for your hospital jobs as a GPST will be the same as your colleagues who are working within that speciality as a hospital ST (although it does also depend on how many years post graduation you are). With each year that you work your basic pay should rise incrementally to reflect that experience; however, the banding supplement that you get will vary from job to job. At present, GPSTs working in general practice get paid a banding supplement of 50%, but this is due to change over the next year to 45%. Again, this is to reflect the change to banding supplements in hospitals, as junior doctors are working fewer hours and more shift work to keep their jobs in line with the European Working Time Directive (EWTD). The argument in general practice is that the hours for GP registrars have not changed (unlike the hospital jobs) and in fact this drop in banding has occurred despite an increase in the amount of extended and out of hours commitments that are now required. In short, your pay in general practice will be equivalent to basic pay relevant to your ST scale plus either 50% or 45%. For a standard trainee doing an ST1 job in general practice this means a gross annual pay of around £45,250, which is still a very reasonable salary compared to your friends in other careers (city bankers aside!). However, this may be decreased further over the next few years as things continue to change, but it'll be a case of wait and see.

What are the main advantages and disadvantages of being a GP?

Everyone will have their own ideas of what does and what doesn't make general practice the career for them and some things that may be advantageous for some, will be a disadvantage to others. Some of the common answers are listed in Table 1.2 but there will be hundreds of others that you come across throughout your career.

In all honesty, the disadvantages are often very minor things that will only bother a few people and for the majority will be outweighed by the advantages, but it's worth mentioning them in an attempt to give a balanced view.

> 'Even though I'd completed my surgical exams and thought I'd end up being a surgeon, I realised that life is too short to spend all of your nights

Table 1.2 Common advantages and disadvantages of general practice

Advantages	Disadvantages
Working autonomously and making your own decisions	Spending the majority of your day working alone and having to make decisions 'off your own back' (the degree of social isolation will vary from practice to practice)
Providing continuity of care for your patients	Dealing with 'heart sink patients' that come to see you on a far too regular basis
Being able to treat existing disease and also get involved in preventing disease and encouraging healthy living	Having to adhere to the Quality and Outcomes Framework (QOF) and code everything that you do
Being able to look after people in the comfort of their own homes	Having to visit patients who live far away from the surgery
Getting to know your patients' families and social circumstances	Having limited access to urgent investigations and results
Being part of the community	Being recognized in the local pub and asked to give advice on your day off!
Ability to chose the amount of sessions worked, making it easier to create a work–life balance	Very rarely getting to 'save lives' in the acute sense of the word (although it can happen even now and then and when it does it can be very satisfying!)
No nights or weekends on call (unless you chose to do out of hours)	You may be required to work in the evenings if your surgery operates an extended hours service

and weekends at work and so changed to GP speciality training and now know the meaning of work–life balance.'

ST2

What personal qualities are required to be a good GP?

This is a difficult one and I'm not sure that there is any one personality type that makes one GP better than another. However, the person specification[2] for the GP application process does give some idea of some of the qualities that are deemed to be particularly important for working in general practice. These are as follows:

- Being empathetic and sensitive—the ability to consider the perspectives of others and treat them with understanding
- Good communication skills—being able to adjust your behaviour and language as appropriate to differing situations
- Being able to think conceptually and solve problems—the ability to 'think outside of the box' and analyse different information and circumstances
- Being able to cope with pressure—the ability to recognize your own limitations and develop ways of coping with pressure
- Being organized—the ability to plan and organize your time effectively
- Being able to manage others and work as a team
- Having professional integrity—having the motivation and ability to take responsibility for your own actions

'I chose general practice because it is a great job offering you the flexibility to be both a generalist and a specialist at the same time, by becoming a GPwSI.'

ST2

REFERENCES

1 Haslam D (2007). *So, You Want to be a GP?* Royal College of General Practitioners, London.
If you'd like a copy of this booklet it can either be downloaded via the RCGP website at http://www.rcgp.org.uk/gp_training/student_forum/becoming_a_gp.aspx or you can request it from the college by emailing careers@rcgp.org.uk. Not only does it give a quick 'day in the life of a GP' excerpt but it also gives a simple breakdown of the path that you would need to take from leaving school to becoming a member of the RCGP. It also discusses different specialities within general practice to consider once you are qualified and gives a brief overview of who the colleges are and what they do. Many of these topics will be covered in more detail in later chapters of this book but the RCGP booklet provides a useful summary.

2 General Practice Training Recruitment. *2010 ST1GP Person Specification Application to enter General Practice Training at ST1*. Available at: http://www.gprecruitment.org.uk/downloads/gp_st1ps.htm (accessed 14 August 2010)

CHAPTER 2

Jumping through the right hoops to get on to the training scheme

Introduction

Most of you reading this book will have hopefully already secured yourself a place on a GP training scheme, but for those who still have this hurdle to jump then this is the most important chapter for you to read. You can treat yourself to the other chapters once you have a formal confirmation letter of your GP training scheme detailing your given rotations.

Despite popular belief by some that general practice is an easy option when it comes to specialist training, VTS placements are in fact relatively difficult to obtain, since not only the arrival of the new selection process, but also as more and more trainees start to realize the merits of general practice training and what it has to offer.

Recruitment on to GP VTS training schemes is organized centrally by the National Recruitment Office for General Practice[1] (NROGP), which can be found online and contains **all** of the essential information that you will need in preparation for the application and assessment process. Unfortunately, however, many private companies have caught on to the increased demand and competition for GP training placements and have been trying to cash in, by charging extortionate amounts for courses and revision books in the promise that it will increase your chances of getting a place. Whilst it may seem tempting to part with your hard earned dosh in the hope that it will improve your chances of success, you really are much better off putting the NROGP website into your computer's favourites and spending a day or two having a good old read through all of the free information available. There are plenty of documents that you can download to give you a step-by-step guide to the application and assessment process and there is also a page dedicated to 'Frequently Asked Questions' which should hopefully answer any other queries that you may have.

Given that I don't want to be tarnished with the same brush as the companies mentioned above, I will therefore keep all of the information in this chapter brief and to the point, but hope that it answers some of the more common questions you may have (and saves you doing too much searching on the website!).

What does GP training involve?

At present, GP training requires you to undertake three years of approved speciality training posts, consisting, on average, of 18 months in general practice and 18 months in hospital posts. In some deaneries there is also the option to apply for an extended four-year scheme which offers you the opportunity to carry out an Academic Clinical Fellowship, where time may be spent either abroad or working on an academic project of interest to you. These are subject to availability and if you are interested then you will need to look at the individual deanery websites for further information.

How many ST1 programmes are available each year?

This varies each year but the anticipated number of ST1 programmes for August 2010 was 3290. The number of programmes per deanery varies from as few as 65 in Northern Ireland to as many as 315 in London.

Is GP recruitment competitive?

If you look at competition ratios by speciality then you'll find it's about average. Data from competition ratios in 2007 showed that there were 6.1 applicants per post for GP training, compared with 0.6 applicants per post for urology and 18 applicants per post for public health[2]. However, more recent data from 2009 suggests that there were approximately two applications for every GPST vacancy at ST1 level[3]. If you're after information on which deaneries are more competitive than others in terms of GP training, then you can look at the competition ratios by deanery on the NROGP website[3]. It does, however, advise you that your choice should be made on where you hope to spend the next three to five years working, rather than where there is less competition for a place. You will also find that competition ratios for each region will vary year on year, as some candidates will be put off applying for the popular deaneries with a subsequent reduction in competition for that year compared with the last.

How long is the training scheme?

At present, GP training is three years, although there is talk that it will change over the next year or so to become five years, in line with the hospital specialities.

Application process

What does the application process involve?

The application process is split into four parts:

Stage 1: Online application to determine eligibility

Stage 2: Assessment 1: Short listing Applied Knowledge Test and Situational Judgement Test

Stage 3: Assessment 2: Selection Assessment Centre (SAC)

Stage 4: Allocation and offer

Who can apply for GP speciality training?

You must have MBBS or an equivalent medical qualification and be eligible to work in the UK. Your eligibility also depends on you having the following:

- Full registration with the GMC at the time of appointment (those requiring PLAB must have successfully completed it within the last three years or have progressed to clinical posts since this time) and hold a current licence to practice[4]
- Evidence of current employment in a UK FPO affiliated Foundation Programme **or** evidence of having achieved the Foundation competences from a UKFPO Foundation Programme or equivalent within the last three years, by time of appointment in line with GMC standards and good medical practice[4]

More detailed information on the application criteria and clinical/personal skills required for GP training can be found on the NROGP website by downloading the '2010 ST1 GP Person Specification'[4] document.

When do I need to apply?

The application process normally opens around the beginning of December for those wanting to start their ST training the following August. Specific dates for each year can be found on the NROGP website, but you should be aware that there is only about a three-week window in which you can apply once the dates are released.

How do I apply?

Stage 1 of the process is the **online application** which determines your eligibility and can **only** be submitted between the designated dates on the website. Applications outside of these designated dates will not be considered and, although it sounds obvious, you are only allowed to submit **one** application. Programmes are available at

ST1 level only, although some deaneries may be offering Academic Clinical Fellowship Programmes as mentioned earlier in the chapter.

You will be required to prove that you have achieved the foundation competencies by providing original evidence, the most appropriate form of which will be determined by specific questions on the application form. Your application will be also assessed against the National Person Specification[4] mentioned earlier, in order to confirm your eligibility and ensure that you have the required clinical and practical skills required, e.g. current valid driving licence and proficient English language skills.

How many deaneries can I apply to?

When submitting your application you are allowed to chose up to four different deaneries, which you should rank in preference order.

Do I have to apply to the same deanery as where I did my Foundation Year jobs?

No, although staying in the same deanery for your ST rotations means that you may have already worked in some of the hospitals in which you may be placed. This is obviously advantageous if you enjoyed your FY1 and FY2 posts and feel familiar with your local hospitals.

Can I apply in advance for a deferred start date?

No, but if you know that you are planning to work abroad for the next training year (e.g. August to August) you MAY be able to take your Stage 2 assessment in the February before you go and carry your result forward to the following year's applications. This saves you having to come back to the UK to sit this component in the middle of your year abroad. It does, however, depend on a number of specific criteria and can only be done in advance for the following year and not any subsequent years. You will nevertheless still be required to submit the actual application on-line at the required time (whilst you are away) and it is only the Stage 2 component that you may be able to do in advance and carry forwards. For more information see the section on the NROGP website entitled 'Working abroad at application time'[5].

The Assessment

What is the Stage 2 Assessment?

This takes the form of two written exams which are taken on the same day by all applicants across the country. Once your application has been accepted you can then proceed to book your place online to sit the Stage 2 Assessment (you will be asked to book this at the nearest available venue to the address on your application form, regardless of which deaneries you have applied to). On arrival at the assessment venue you will be expected to provide photographic identification and will not be able to sit the exam unless you can do so.

The assessment consists of two machine-marked papers:

Paper one: Clinical Problem Solving

This paper is similar to the Applied Knowledge Test (AKT) that you will sit as part of your RCGP membership exams at the end of your GP training, but is based at FY2 knowledge levels. It lasts 90 minutes and consists of multiple choice questions based around clinical scenarios. The aim is to test your judgement and problem solving skills plus your ability to apply your knowledge in the diagnosis and management of common conditions. Because the paper is machine marked, you will complete your answers on a separate sheet by filling in lozenges based on your selected answer.

Paper two: Professional Dilemmas

This paper lasts 120 minutes and presents you with different professional dilemmas that you may come across when practising as a doctor and assesses how you would deal with them. It is a chance to show evidence of your professional integrity plus demonstrate your ability to show sympathy, empathy, and how you cope with pressure. You will be scored depending on how close your responses are to those given by a panel of expert GPs.

Neither paper is negatively marked, so make sure that you give an answer for every question and as always, that you read the questions carefully.

Will I have to sit a written exam?

Yes, this is Stage 2 of the assessment. See previous question **'WHAT IS THE STAGE 2 ASSESSMENT?'**.

What is the best way to prepare for the Stage 2 Assessment?

The best preparation is to make sure that you have fully read all of the information available on the NROGP website and have looked through examples of what kind of questions or scenarios you may be faced with. If you feel that you would like to do some clinical preparation then the best advice I can give is to have a flick through a general practice handbook. This will enable you to brush up on some of the specialities that you may be a little rusty on, without taking up too much of your time or scaring you witless about things that you don't need to know.

> 'In preparing for our Stage 2 Assessment, my friends and I found it really useful to go through the prioritization exercises together, so that we could explain the reasoning behind our rankings. It also allowed us to just 'shed a different light' on things! The other thing we found useful was practising

the group discussions together—it enabled us to look at the discussion from lots of different angles which definitely helped in the real exam'.

FY2

What happens after the Stage 2 Assessment?

If you successfully complete the Stage 2 Assessment (which relies on you performing adequately in **both** papers), then you will be sent an email inviting you to book your place online for the Selection Assessment Centre (SAC), usually at your first choice deanery. This is referred to as **Stage 3** of the application process and involves candidates completing three exercises which are observed and assessed by trained assessors. The assessors will be looking to identify which candidates demonstrate the competences listed in the person specification document[4].

The three exercises are as follows:

- **Patient simulation exercise**—this involves a role play scenario with a simulated patient and clinical case. You will not be expected to examine the patient and it shouldn't involve anything that a doctor at your stage of training couldn't deal with.
- **Written exercise**—this is an exercise which has no specific right or wrong answers but looks at your judgement skills. For example, it may look at your prioritization skills by asking you to rank the order in which you would deal with five possible hospital setting scenarios and explain your reasoning for the answers you have given.
- **Group exercise**—after being randomly allocated to a group of other candidates sitting the SAC, you will be asked to discuss a series of difficult scenarios within your group, the whole of which will be observed by an assessor.

The Stage 3 process looks specifically at areas such as professional integrity, communication skills, listening, empathy and sensitivity, team working, decision-making, approaches to learning and updating, and fitness to practice. You might see some common themes arising such as:

- GMC guidance on the duties of a doctor[6]
- RCGP guidance on good medical practice for GPs[7]
- Revalidation[8]

For further information and examples of the types of scenarios used in the SAC, see 'Example Scenarios from Selection Centre Exercises'[9] on the NROGP website.

Will I have a formal interview?

No, not in the true sense of the word. See previous question **'WHAT HAPPENS AFTER THE STAGE 2 ASSESSMENT?'**

What will I need to take with me on the day of the SAC?

THE SAC CHECKLIST

- References (supplied on the Structured Reference Form[10] as requested)
- Photographic proof of identity, e.g. passport or driving licence
- Proof of right to work in the UK
- Original GMC certificate plus a photocopy
- Original medical qualification certificate and photocopy
- Original copies of evidence of having completed the Foundation Years' competences
- Valid UK driving licence (or evidence of suitable arrangements that you have made to enable you to provide emergency or domiciliary care)

Offer and scheme allocation

How will I know if I have been successful in the Stage 3 Assessment?

Offers of GP placements will be emailed to successful candidates as soon as possible after the SAC. Some deaneries will provide information on your allocated posts, whereas others will just confirm that you have a place on the scheme, and will allocate the posts once you have officially confirmed your offer. You have 48 hours in which to respond to the offer otherwise it will be withdrawn and offered to another candidate. Only **one** GP training programme will be offered to each candidate and if this is declined then that candidate will not be offered any others. It's also important to mention that once you have accepted your place on the GP training scheme, then you are expected to withdraw from other speciality applications that may be active (unless you are awaiting the result of an application to Core Medical Training (CMT), psychiatry, or paediatrics, in which case you may then **hold** your offer until the result from that speciality is known).

What if I don't get offered a place?

Candidates who are considered suitable for GP training but are not ranked highly enough to be offered a place at their first choice deanery will be classed as 'reserve candidates' and will be considered on a ranked basis for any posts that are declined within that deanery (local clearing). National clearing will also take place in order to fill those remaining vacancies in other deaneries and will again be done based on each candidate's overall ranking rather than their original deanery choices.

REFERENCES

1 The National Recruitment Office for General Practice Training. Available at: http://www.gprecruitment.org.uk/ (accessed 14 August 2010).

2 Modernising Medical Careers. *Competition Ratios 2007 ST1*. Available at: http://www.mmc.nhs.uk/docs/ST1_specialtysummary.pdf (accessed 14 August 2010).

3 General Practice Training Recruitment. *Competition Ratios*. Available at: http://www.gprecruitment.org.uk/recruitment/competition_ratios.htm (accessed 14 August 2009).

4 General Practice Training Recruitment. *2010 ST1GP Person Specification. Application to enter General Practice Training at ST1*. Available at: http://www.gprecruitment.org.uk/downloads/gp_st1ps.htm (accessed 14 August 2010).

5 General Practice Training Recruitment. *Working abroad at application time*. Available at: http://www.gprecruitment.org.uk/abroad.htm (accessed 14 August 2010).

6 General Medical Council. *Good Medical Practice*. Available at: http://www.gmc-uk.org/guidance/good_medical_practice.asp (accessed 14 August 2010).

7 RCGP. *Good Medical Practice for General Practitioners*. Available at: http://www.rcgp.org.uk/PDF/PDS_Good_Medical_Practice_for_GPs_July_2008.pdf (accessed 14 August 2010).

8 RCGP. *Guide to revalidation for GPs*. Available at: http://www.rcgp.org.uk/PDF/PDS_Guide_to_Revalidation_for_GPs.pdf (accessed 14 August 2010).

9 General Practice Training Recruitment. *Example Scenarios from Selection Centre Exercises*. Available at: http://www.gprecruitment.org.uk/downloads/gp_ex2.pdf (accessed 14 August 2010).

10 General Practice Recruitment. *Clinical Structured Reference Form*. Available at: http://www.gprecruitment.org.uk/downloads/gp_srf.htm (accessed 14 August 2010).

PART 2

Early ST1

CHAPTER 3

You've got a training post—congratulations! Now what?

Firstly let me start by saying congratulations on finally securing yourself a GP training post! No doubt it will have been a lengthy process to get this far, however the next few years will make it all worthwhile by bringing you lots of new challenges and excitement, not to mention the title of MRCGP at the end of it all if you are successful. Whilst you probably have several months left in hospital posts before you start your journey along the road to becoming a GP, there are several things to think about and probably many questions that you may want answered before you start. This chapter aims to answer some of the frequent queries that trainees often have and also hopes to give you an insight into what things to prepare in advance.

About the rotations

What is my NTN?

This is your National Training Number and should be detailed on your rotation offer letter. Make sure that you keep this number safe as it may be required for various different purposes at a later date (when your offer letter has been long gone via the recycling bin!).

Tip: Make sure that you keep your NTN safe as it may be required for various different purposes at a later date (when your offer letter has been long gone via the recycling bin!).

What posts may I be allocated?

This will vary depending on what posts you have completed previously and also what is available within your area. Some deaneries will randomly allocate rotations, whereas others may look at your previous experience and try not to duplicate specialities in which you have already worked. Most rotations will incorporate at least one of the core areas such as medicine or A&E, plus offer the opportunity to work in some of the more specialized posts such as O&G, ENT, paediatrics, and psychiatry. There are also some newer posts becoming available within certain deaneries that allow you to work in fields such as public health or palliative care, so don't be surprised if you have one of these included within your rotation.

Does the rotation that I do affect my pay?

In short, yes. As with your previous hospital jobs, the pay that you receive whilst doing a hospital or speciality post will depend on the hours that you do and how unsociable they are. For example, if you are posted in public health, your working hours are likely to be between nine and five with no weekends, so your pay will reflect this and is likely to attract a smaller banding supplement. However, if you're posted in an O&G job with a one in three on-call rota covering night and weekend shifts then you are likely to get a fairly high banding supplement and hence overall a much higher rate of pay. Essentially you don't get something for nothing and whilst it's frustrating that the amount of money that you take home is likely to change every six months, as long as you prepare for this in advance it shouldn't be too much of a problem.

Can I change my posts once they have been allocated?

This depends on how accommodating your training programme directors (TPDs) are and what your deanery will allow. Trying to change your posts can cause problems when you come to apply to PMETB at the end of your training, as if you haven't covered a broad enough spectrum of specialities then they may not agree to your certification. Changing posts is possible but is often very stressful and confusing for everyone involved. Some deaneries will allow you to swap a rotation if you can find a willing colleague to do this with but that certainly won't be the case for most deaneries and you may be better off just grinning and bearing it. Repeating a speciality can never be a bad thing as there is bound to be plenty of knowledge that you still have yet to learn and sometimes life is just too short to complicate things by trying to swap. However, if you really are unhappy and want to consider swapping your rotations then you have nothing to lose by approaching your TPDs and just asking (although try not to stalk them by sending emails about it every other day as this is likely to just send them insane! Remember that they are working GPs too and have plenty of other things to try and squeeze into their busy lives).

How many different posts will I do?

At the time of this book going to print the GP training scheme was a three-year programme that included around 18 months of general practice experience plus 18 months of hospital or alternative speciality experience. Given that each post lasts for six months (if you are working full time) you will get an additional three posts to your GP placements. This does, however, vary between each deanery and it is therefore worth checking the individual websites to see what their arrangements are.

How long will each post last?

Most posts including those in general practice are now allocated for six months, but this will vary if you are working at less than full time. Six months is deemed to be a sufficient amount of time for you to get a grasp of the speciality and ensure that you cover the necessary curriculum statement areas (although this relies on you making the most of the opportunities that each post offers).

Will I get any GP experience in the first two years?

Yes, but the amount that you get will vary depending on your rotation and where you work.

Do I get to choose where I do my GP component?

Possibly yes, but it depends on the area in which you work. If you do get the opportunity to choose then it's important to start thinking early on about where you would like to spend your time. Make sure that you consider your options wisely and if possible try to visit different surgeries to get a feel for the type of place where you would like to work. Some schemes put on an event where all of the training practices within the area set up their own stand to 'tout for business' (as it were) and provide trainees with further information on what experiences their practice can offer. You may also want to think about choosing a practice for ST3 that offers something different to that of the practice where you may be working in the earlier training years. Getting a broad experience of different types of practices and the way in which they work will equip you well for when you are considering job opportunities at the end of your training. Whilst it seems a long way off now, it won't be long before you're searching through job adverts to find a practice where you would like to work on a long-term basis.

Tip: Getting a broad experience of different types of practices and the way in which they work will equip you well for when you are considering job opportunities at the end of your training.

Essentially all training practices should offer similar experiences as they are carefully regulated by the deanery, but it's important that you find somewhere that you think you will enjoy working and will have a trainer that you can feel comfortable and work well with. Remember, for your ST3 job you will be based at the same surgery for a full year, which is likely to be the longest amount of time that you will have spent within the same post since you started your postgraduate training. Whilst a year can fly by if you are having fun, it can also feel like an age if you are somewhere that you are not enjoying. You obviously can't predict what it will be like to work in each practice but there is a lot to be said for that 'feel' that you get when you walk in to the surgery. First impressions really can count.

Will there be any formal teaching or study days?

Each deanery will plan their study days and teaching in different ways. Some may have a series of intensive study courses over two to three days a few times throughout the year, whilst others will have a specific half-day release system (one afternoon a week) where all trainees within that region meet for formal teaching. Find out early on what your local training scheme offers and plan ahead, so that you can limit the difficulties you may have trying to get released from your different posts. Some specialities are renowned for not enabling GP trainees to get to their designated teaching due to ward commitments and on-call duties. For further information on what study days and teaching may be available within your area see **CHAPTER 9,'HOW TO GET THE BEST OUT OF YOUR REGIONAL VTS TEACHING'**.

Tip: In terms of VTS teaching, find out early on what your local training scheme offers and plan ahead, so that you can limit the difficulties you may have trying to get released from your different posts.

Where can I find out more information on my regional VTS scheme before I start the rotation?

The best place to start looking is your regional VTS website. Most areas will have one, although the content will vary hugely from one area to another. The information available ranges from a schedule of forthcoming teaching sessions to information on training practices and which surgeries have vacancies for ST placements.

Approaching current trainees is probably the best way of finding out all that you need to know about a practice and many VTS scheme now operate a 'buddy system' whereby each new trainee is assigned to a current GPST so that they have someone to approach for common questions and queries about the training programme. You may also be allocated one of your course organizers as a mentor at the start of your training.

Some areas will also have their own Google Groups site (or other form of networking/blog site) where people can post information, queries, or general chit chat and social announcements. If this exists in your area then it is often a good place to look to find out more information, as the chances are someone else will have had the same query as you and therefore the answer may be already there. If you do have a web-based facility such as this available to you, make sure that you know how to register for access as it's a good way of keeping up to date with what other trainees are up to. If you don't have a blog site for your VTS scheme, perhaps think about creating one. It's relatively simple to do and your colleagues will definitely thank you for it. What's more, it's something else that you could put on your ePortfolio to prove your commitment in terms of contributing towards general practice training.

Tip: If you don't have a blog site for your VTS scheme, perhaps think about creating one. It's relatively simple to do and your colleagues will definitely thank you for it.

Who's who on the VTS scheme?

Within each deanery, the training programmes are divided into separate VTS schemes based on region within that area. Each VTS scheme is usually run by a number of Training Programme Directors (TPDs) or Postgraduate Clinical Medical Educators (previously entitled 'course organizers'), who are practising GPs with a special interest in medical education and teaching. The TPDs are responsible for developing and delivering your local training programme and should also provide pastoral support to trainees in difficulty. These people should be your first port of call should you have any queries or issues that cannot be answered or resolved by your colleagues. Many of these TPDs also sit on the ARCP panel so they can be useful people to make friends with early on. Each scheme is overseen by the Area Director who is responsible for organizing trainees' rotations and coordinating the TPDs. The Area Director also provides the interface between the individual VTS schemes and the associated deanery.

What is the RCGP curriculum?

This identifies the knowledge, competencies, clinical/communicational skills, and professional attitudes required of a doctor intending to undertake independent practice in the UK as a GP. It forms the skeleton of the assessment processes that take place during your GP training years and is an essential read for all new trainees. It is split into many topic areas which include not only clinical subjects but also administrative and managerial ones too. It basically provides a summary of what you are expected to learn and achieve during your speciality training so can also be a useful guide when it comes to revision for exams.

The paperwork stuff

What paperwork do I need to fill in before I start?

The most important thing you need to do, regardless of which posts you are allocated, is to register with the RCGP so that you can gain access to the all-essential ePortfolio. Registering with the College will unfortunately cost you money but there's no getting around it and the sooner you do it, the sooner you can start contributing to your ePortfolio.

How do I register with the RCGP?

Registering with the college is very straightforward and can be done online via the RCGP website under the section headed 'New professionals – Associates in Training (AiTs)'. There are several different types of AiT package available depending on whether you join the scheme in ST1, ST2, or ST3 (although from 2009 all trainees should be entering at ST1 level). You will be required to pay an initial one-off registration fee of £142 (based on ST1 rates) followed by an annual subscription of £213 which is paid via direct debit on the first of April each year. The total paid over the three years works out at £781 and entitles you to ePortfolio access, discounts on exams and courses, access to online learning and self-assessment, 10% off the RCGP bookshop, and subscription to both the *BJGP* and the AiT journal *InnovAiT*. You can also register solely for use of the ePortfolio without the benefits of becoming an AiT for £567, although by the time you have factored in the extra you will pay for courses, exams, and books, it isn't really worth it. Missing out on receiving the *InnovAiT* journal is also not ideal, as it contains a large amount of curriculum-relevant clinical knowledge not to mention other additional benefits such as AKT practice questions.

For those who will be applying for accreditation via Article 11 the situation is slightly different and subject to change; however, up-to-date information can be found on the RCGP website.

If you have any other general enquiries about registering as an AiT, further details can be obtained via the RCGP AiT helpline at ait@rcgp.org.uk or on 020 7344 3078.

What is Article 11?

Article 11 is the way in which trainees can apply for GP certification if they have completed posts either outside the UK or outside of a GP training scheme, that they wish to contribute towards their final assessment. It is also the required route for those who have completed their posts more than seven years before they apply for certification.

What is *InnovAiT*?

This a journal produced by Oxford University Press in conjunction with the RCGP for associates in training. It is produced monthly and provides both clinical knowledge and practical tips on general practice training, plus sample AKT questions and relevant news articles. It is based around the GP curriculum so is an invaluable resource when it comes to exams and focusing your learning. Many of the articles are written by trainees and therefore provide knowledge which is not only pitched at the right level of detail but is relevant to what trainees want and need to know. Trainees who register as AiTs will receive this free as part of their membership but for those who are not eligible it can be purchased as a monthly subscription via Oxford University Press or the RCGP.

Should I join the BMA (if I haven't done so already)?

Whether or not you join the BMA is entirely up to you although many of you will have already done this. It's worth knowing that if you are a member and currently receive the BMA journal, you may want to consider contacting the membership department to request that you receive the GP version of the journal. Whilst this might not be so important in the early stages of your training, you will need to do this once you start thinking about applying for jobs, as most are advertised in the GP version of the *BMJ*. You can of course also access this online so it isn't essential; however, being a BMA member is worth a great deal more than just receiving the journal. Again this may become more apparent when you are applying for jobs and require their contract checking service, but they can also be of use to you during the training years when you may need them to help you fight further pay supplement cuts or other training issues. They in theory act as your 'union' and can therefore be used as a negotiating body if you have any problems or disputes.

As a BMA member you are also entitled to a discount on some exam revision websites and medical equipment that you may require for your doctor's bag.

Do I need to change my GMC registration in any way?

No, although you will need to update your GMC registration once you complete your training and get your CCT.

How will my training affect my medical defence organization membership?

Whilst you are in hospital posts your medical defence organization membership will remain the same. However, when you move in to a general practice post you will need to inform your medical defence organization and provide them with the exact start and finish dates of your time spent in general practice so that they can calculate your new premium. This costs **considerably more** whilst in general practice but can be claimed back via your local Primary Care Trust (PCT). You will, however, be expected to pay the hospital proportion of the membership and it is only the additional amount required for general practice that can be claimed back.

> Tip: Reading the relevant specialty chapter in the *Oxford Handbook of General Practice* before starting your hospital posts should prevent you from turning up clueless on day one!

What is Form R?

This is a form which must be completed annually by each trainee to remain registered for speciality training with your deanery. It should not be confused with an R7 form which is completely different. At the start of your training you will be requested to complete a Form R and mail it to the relevant Postgraduate Dean. Reminders to complete this in subsequent years are normally sent via your ePortfolio.

What is an R7?

An R7 is a contract between you, the GP training practice, and the deanery. It exists to notify the deanery of your placement in a training practice so that your pay can be correctly calculated. Without it, you cannot be paid, as your surgery will only receive reimbursement from the PCT once an R7 is in place. It is different to Form R and is not a contract of employment. The contract of employment for your general practice posts is signed via your ePortfolio.

Other things to think about

What books should I buy in advance?

The most useful book that you can have on your shelf is the *Oxford Handbook of General Practice*. As with all of the other Oxford Handbooks it is a compact useful reference guide and covers all of the important aspects of general practice, plus clinical information for each speciality at the appropriate level. You can't go far wrong if you have one of these handy whilst at work and they also act as a useful revision reference guide when it comes to exam time. Reading the relevant specialty chapter before starting your hospital posts should also mean that you don't turn up clueless on day one.

Tip: Don't forget to change your role with your medical defence organization prior to starting any GP placements—it costs considerably more and whilst it can be claimed back from the PCT at a later date, you will be required to pay it up front!

For those who are up to date with technology you can also download a version of this book on to your PDA so that you can quickly look up information on the run. Both the book and the PDA version are similar in price and are available from all good bookstores (although as a RCGP AiT you will get 10% discount on their bookstore).

Are there any helpful websites I should be aware of?

Yes. By far the most important website which you need to be able to navigate is that of the RCGP. It can be found at http://www.rcgp.org.uk and contains a wealth of knowledge and information about every aspect of GP training and assessment. Being able to find your way around the website will enable you to gain maximum benefit from the college in terms of support, administrative information, clinical assessment, and practical knowledge. It may sound like an obvious thing to get to grips with but you'll be amazed how many trainees are unaware of what services the RCGP website offers. At this early stage, one of the most crucial areas of the website to be familiar with is the GP curriculum documents, which provide the skeleton on which GP training is built.

One of the other most useful websites is http://www.gpnotebook.co.uk. In a similar way to the *Oxford Handbook of General Practice* this website offers a quick reference guide to most clinical aspects of general practice and more importantly is completely free to use (at present). As there aren't many places without internet access these days, it means that wherever you are you can quickly look up signs, symptoms or diagnoses to find out further clinical information if you ever get stuck.

Another website which you may find useful is http://www.primarycareforms.com. This website was created by a GP trainee who wanted to solve the difficulties faced by all primary care providers when trying to hunt down the correct forms for various different clinical and non-clinical purposes. It contains a huge amount of links which enable the user to find and download the forms that they require in one easy step, potentially saving hours of time searching the internet.

When should I start thinking about exams?

Given that you probably haven't even started your rotation yet, it may seem a little premature to start thinking about your forthcoming exams at this stage. However, not only are the MRCGP exams expensive (and therefore may require considerable financial planning) but you may want to think in advance about other exams and courses that you wish to do, so that you can plan your time (and study leave) accordingly.

Mapping out a brief idea of what you want to achieve over the next three years of your GP training will help to focus your learning and proportion your time appropriately. Deciding at the last minute that you would like to invest in doing some extra diplomas before you qualify is likely to be stressful, expensive, and distract you from your ultimate goal which is successfully completing the MRCGP exams.

I've heard a lot in the news recently about revalidation. When do I need to start thinking about this?

It has been made very clear that revalidation is starting and will form an essential part of our training and future careers. Whilst the ePortfolio should provide an extensive log of what you are experiencing within your training, it may be worth starting your own revalidation folder from day one. Use this folder to store all of your important documents and certificates, plus any 'thank you' cards you receive from patients or relatives. Include in your folder paper copies of any multi-source feedback (MSF) summaries or appraisal meetings, printed copies of any audit projects or significant event meetings, plus a record of your hepatitis B status and medical defence organization membership. This folder can then be adapted at a later stage to become useful when applying for jobs, working as a locum, and more importantly helping you to prove what you've been up to over the last three years when it comes to appraisal and revalidation.

CHAPTER 4

What lies ahead? Introduction to the assessment process

General information

Now that you have been successful in getting a place on the training scheme, you'll probably be keen to know and understand the various elements of the assessment process that should eventually lead to you attaining your MRCGP. Since 2007 there has been a single training and assessment system for UK-trained doctors wishing to obtain a CCT (Certificate of Completion of Training) in General Practice, satisfactory completion of which is an essential requirement for entry to the General Medical Council's (GMC's) GP register and for membership of the Royal College of General Practitioners[1].

This chapter aims to guide you through the assessment process and hopefully answer any queries that you may have along the way.

How is the MRCGP assessed?

There are three components to the MRCGP assessment:

- Workplace-Based Assessment (WPBA)
- Applied Knowledge Test (AKT)
- Clinical Skills Assessment (CSA)

WPBA will be discussed in more detail within this chapter, but for further information on the AKT and CSA see **CHAPTERS 14 AND 18** accordingly. In addition to the above components, you are required to create a continuous learning log and PDP via your ePortfolio, attend a CPR update course (and achieve the relevant certification), and complete a total of at least 72 hours of out of hours experience. More information on the learning log and PDP can be found in **CHAPTER 5, 'THE EPORTFOLIO EXPLAINED'**. All of these components of the MRCGP are then tracked via the 'Progress to Certification' section of your ePortfolio.

Who is responsible for the different parts of the MRCGP assessment?

The RCGP is responsible for the AKT and CSA whilst the deanery is responsible for the delivery of the WPBA. Every year your WPBA will be evaluated by the Annual Review of Competence Progression (ARCP) panel within your deanery, who will then make a judgement as to whether you are meeting the required competencies at the expected rate and can therefore be entitled to progress onto the next stage of the ST programme. This judgement is based on the evidence that you have collected within your ePortfolio and it is your responsibility to make sure that this evidence is collected in plenty of time for each review.

How do I get the CPR section signed off?

All trainees must demonstrate their competence in CPR and AED usage by attending a relevant course or assessment. This can be achieved by either attending the CPR training day within your surgery (which usually occurs every 18 months) or getting a group of trainees together and organizing for the local hospital resuscitation officer to provide a training session for you. Once completed you will need to submit a valid certificate (signed by a qualified ALS instructor or equivalent) via your learning log. You will not need to attend a full ALS course as you will not be expected to be able to use advanced airway methods or administer drugs. Each update is valid for three years and you will need to pay for this yourself, although you may be able to claim it back via your study budget allowance.

WPBA

WPBA—what is it?

WPBA is defined as the 'evaluation of a doctor's progress over time in their performance in those areas of professional practice best tested in the workplace. It is a process through which evidence of competence in independent practice is gathered in a structured and systematic framework'[2]. Put more simply, it is a method of collecting evidence in order to prove (alongside the AKT and CSA) that at the end of your three years of training you are able to practice independently as a GP. It is designed to be a developmental process by which you accumulate assessments along the way. These assessments are then analysed prior to each six-monthly review with your educational supervisor (ES), to ensure that you are meeting the required expectations for each stage of your training.

What does the WPBA involve?

Essentially it involves a framework of 12 professional competency areas in which you are expected to be proficient by the end of your training. The evidence to prove your ability in these areas is collected using seven different validated tools which are as follows:

- Case-based discussion (CbD)
- Consultation observation tool (COT—primary care only)
- Multi-source feedback (MSF)
- Patient satisfaction questionnaire (PSQ—primary care only)
- Clinical evaluation exercise (mini-CEX—hospital posts only)
- Direct observation of procedural skills (DOPS)
- Clinical supervisor's report

Completion of these tools allows your training to be continually assessed throughout the three years and enables your ES to make a qualitative judgement on your progress at six-monthly intervals. In essence, these tools form part of a formative assessment and there is therefore no pass or fail aspect to them. Instead, your ability in each submitted tool will be classified as one of the following:

- Insufficient evidence
- Needs further development
- Competent
- Excellent

It is unlikely that you will be achieving more than 'Needs further development' in the early stages of your training, as it would be unrealistic to expect you to have the proficiency of an independent GP at this stage (which is the level against which your capability is judged). Whilst this can seem disheartening at first, remember that your WPBA needs to show evidence of progress and therefore it probably won't be until the latter half of your training that you start to achieve 'competent' or even 'excellent' in the required professional competency areas.

How much weight does the WPBA carry?

There is no specific weighting to each component of the MRCGP assessment, as you will need to successfully complete all three in order to achieve the qualification. The WPBA (unlike the AKT and CSA) is more of a formative assessment, however you are still required to demonstrate proficiency in all of the 12 competency areas in order to 'pass' this component of the membership.

Does anybody fail their WPBA?

As stated previously, there is no pass or fail element to the WPBA although you will need to have gained all of the required competencies in order to complete your training. If you have not managed to achieve this by the end of the three years, you will need to apply to extend your training in order to allow you more time to do so. This request will not automatically be granted and will be at the discretion of your deanery. However, with careful planning, determination, and hard work, there is no reason why anyone shouldn't be able to successfully complete all the required elements within the training time allocated.

Evidence

CbDs

WHAT IS A CBD?

These are structured interviews which take place between a trainee and their educational supervisor (trainer) to assess that trainee's judgement in a variety of different cases. The cases are chosen by the trainee and presented to the trainer in advance, so that both parties are able to make sufficient preparation in order for the discussion to take place smoothly. It also enables the trainer to decide which competency areas the chosen cases highlight and can therefore be assessed. In ST1 and ST2, the trainee is expected to prepare two different clinical cases of which the trainer will chose one to discuss. In ST3 the trainee is expected to choose four cases to prepare of which one or two may be discussed. It is not expected that the trainer will be able to discuss all competency areas within each case and instead two or three may be used as the focus. Think carefully about the cases that you select, as over the course of your training the aim is to cover and become proficient in all of the 12 competency areas. This may require you to 'hunt down' specific cases which you feel will demonstrate your abilities in certain areas.

As previously discussed, don't be disheartened if your trainer classifies you as showing 'insufficient evidence' or needing 'further development', as this is perfectly normal in the early stages and will help you to determine the further learning needs that you may have.

> Tip: Think carefully about the cases that you select, as over the course of your training, the aim is to cover and become proficient in all of the 12 competency areas.

> Tip: Don't be disheartened if your trainer classifies you as showing 'insufficient evidence' or needing 'further development', as this is perfectly normal in the early stages and will help you to determine the further learning needs that you may have.

WHEN MUST I START DOING CBDS?

As with all of the assessment tools within the WPBA, completing CbDs must start from day one of your training. Make sure that you are aware of how many you are expected to have completed within each post and plan ahead accordingly. Trying to get them done in the last two weeks of your post is not advisable and can be particularly stressful for everyone involved. Discussing cases with your seniors is likely to be something that you are doing on a fairly regular basis anyway, so planning them in advance and formalizing the process is not only good for your learning but enables you to make sure that you are keeping up with the required number of CbDs.

> Tip: Make sure that you are aware of how many of each assessment tool you are expected to have completed within each post and plan ahead accordingly.

HOW MANY CBDS DO I HAVE TO DO?

Details of how many of each assessment tool you must complete at each stage of your training can be found in Table 4.1. In order to find out how many you have completed, you will need to check the 'Evidence' section of your ePortfolio. Once the minimum number required for that period of review have been submitted (assuming that you have selected the appropriate stage and level of your training in the drop-down list)

Table 4.1 Minimum requirements for WPBA assessment tool completion at each stage of training

	Month	COT/ mini-CEX	CbD	MSF	PSQ	DOPS	CSR
ST1	6	3	3	1[a]		As appropriate	1[c]
ST1	12	3	3	1[a]		As appropriate	1[c]
ST2	18	3	3		1[b]	As appropriate	1[c]
ST2	24	3	3		1[b]	As appropriate	1[c]
ST3	30	6	6	1[d]	0	As appropriate	0
ST3	34	6	6	1[d]	1	As appropriate	0

[a] with five clinicians only
[b] if in primary care
[c] at the end of each 6-month post regardless of whether hospital or GP
[d] 5 clinicians and 5 non-clinicians

then the number will change from red to green. Any assessments done in addition to the minimum amount will only count towards the forthcoming review and cannot be carried forward to subsequent or future reviews (i.e. just because you manage to do 10 assessments in one six-month period does not mean that you can do fewer than the minimum amount in the next). It's worth pointing out at this stage that most good trainees will do more than the minimum number required for each review period, as not only does this show commitment when it comes to ARCP review but it also aids their learning and overall development.

Less than full time trainees or those who have had absences or extended training will have different requirements; the details of which can be found on the RCGP website.

> 'Aim to do more CbDs/CEXs/COTs than recommended—this gives you more spread of curriculum coverage and better performances.'
>
> ST2

COTs

WHAT ARE COTS?

These assessments are carried out in primary care and require your trainer to observe a clinical interaction between you and a patient. This can either be achieved by collecting a series of videoed consultations, having your trainer sit in on your surgery, or both. Some training practices set aside specific clinics for these to be carried out, which enable you as the trainee to have more time to negotiate the video camera or have a feedback session with your trainer afterwards.

> Tip: Try and have your trainer sit in on your consultations **as well as** recording videos for your COTs. The dynamics of the consultation will differ with each method and the more feedback you can get the better!

HOW MANY COTS DO I HAVE TO DO?

Details of how many of each assessment tool you must complete at each stage of your training can be found in Table 4.1. In order to check how many COTs you have completed, you will need to check the 'Evidence' section of your ePortfolio. Once you have submitted the minimum number required for that period (assuming that you have selected the appropriate ST year and review from the drop-down list) the number will change from red to green.

WHAT IS THE DIFFERENCE BETWEEN A MINI-CEX AND A COT?

Mini-CEXs are the secondary care equivalent to COTs and involve the trainee being observed during a doctor–patient interaction. Each mini-CEX should represent a different clinical problem and should be assessed by a different observer on each occasion[3]. Immediate feedback should be provided by the observer and recorded in the trainee's ePortfolio.

DOPS

WHAT ARE DOPS?

The concept of DOPS will be very familiar to those who have completed the foundation years training in the UK; however, for those who haven't, they are in basic terms a way of assessing your ability in a number of practical procedural skills. There are eight mandatory procedures which have been selected as sufficiently important and/or technically demanding to warrant specific assessment[4]. These are as follows:

- Breast examination
- Female genital examination
- Male genital examination
- Rectal examination
- Prostate examination
- Cervical cytology
- Testing for blood glucose
- Application of simple dressings

Eleven optional and 11 foundation DOPS can also be submitted should you get the opportunity to do so, although these do not form part of the WPBA. Each DOPS assessment must be submitted by someone who has seen you perform the skill first hand during your GP training rotation. The 'DOPS present' column then shows the number of DOPS submitted for each particular skill and allows you to view the entries by clicking on the relevant number. You must continue submitting assessments for each of the eight mandatory DOPS until you have reached the level of 'meets expectations' or above in that skill. Once this has been achieved you are permitted to update your personal rating of each of these skills to 'competent'.

HOW MANY DOPS DO I NEED TO GET DONE AT EACH STAGE?

Unlike the other WPBA assessment tools, there is no specific number of DOPS assessments that you are required to do by the end of each post. Instead, you should be completing them when you have the opportunity, be that either in hospital or primary care. However, you should ideally aim to get your DOPS assessments completed during the first two years of your training (and preferably within your hospital posts as it can be slightly more difficult to observe such procedures within a general practice setting where you work predominantly on your own). If you do still have outstanding DOPS at the start of your ST3 year then make sure that you discuss this with your trainer, so that they understand why during their morning surgery you might request that they watch you examine someone's rectum!

WHO CAN OBSERVE MY DOPS?

These assessments can be carried out by experienced SpRs, staff grades, nursing staff, consultants, or GPs.

HOW DO I COMPLETE A DOPS?

You will need to find an appropriate observer, the necessary equipment, and a willing patient. Each DOPS assessment including feedback should take around 20 minutes and once completed should be submitted onto your ePortfolio by the observer. For further information on how to achieve this see **CHAPTER 5, 'THE EPORTFOLIO EXPLAINED'**.

WHAT ARE THE OPTIONAL DOPS?

These are skills which you may have the opportunity to carry out in your speciality posts and whilst they are not mandatory for your general practice training, they are worth completing, as they may become an important addition to your CV at a later date.

DO I NEED TO REPEAT THE FOUNDATION DOPS?

You should not routinely be expected to repeat the foundation DOPS although should your ES feel that your ability in these areas is questionable then they may request that you do so.

I DID MOST OF THE MANDATORY DOPS DURING MY FOUNDATION YEARS—CAN I CARRY THEM OVER TO MY GP EPORTFOLIO?

No. All mandatory DOPS must be carried out whilst in general practice training, even if they have already been achieved in previous posts.

I HAVE DOPS FOR THE FOUNDATION SKILLS ON PAPER—HOW DO I ADD THEM?

This can be done within your skills log by uploading the documents into your ePortfolio. The paper copies will obviously need to have first been scanned and saved to your computer in order for you to do this. If you don't have the technical equipment available for this to be possible, then don't worry, as you are not expected to provide evidence of competency in the foundation skills DOPS unless your trainer has specifically asked you to do so.

I'M STRUGGLING TO GET THE INTIMATE EXAMINATIONS SIGNED OFF—HELP!

This can be a common problem, in particular for male trainees who do not have a gynaecology post within their training. The same is true for females who do not get the opportunity to do a urology or general surgery post. For this reason it is extremely important that you think in advance about how you are going to achieve your DOPS over the course of your training and try as much as possible to get them signed off during your hospital posts (where patients tend to be less choosy about the sex of the doctor that they see!). Very few women will choose to see an unfamiliar male GP about a genital problem and vice versa. If you are concerned that you may struggle in these areas, then discuss this with your educational supervisor early on and see whether you could organize to attend a clinic, where you may get the required experience and subsequent observation.

> Tip: Think in advance about how you are going to achieve your DOPS over the course of your training and try as much as possible to get them signed off during your hospital posts.

PSQ

WHAT IS THE PSQ?

The PSQ provides feedback for trainees on how their communication skills are perceived by the patient during the consultation. It looks particularly at your ability to put the patient at ease, listen to and understand their concerns, show compassion, and explain to them their condition and share how it will be managed. You will need to complete at least 40 of these questionnaires for each PSQ submission.

HOW DO I ORGANIZE MY PSQ?

Firstly you will need to print off at least 45–50 copies of the required form (details of how to do this can be found in **CHAPTER 5, 'THE EPORTFOLIO EXPLAINED'**). These should then be given to the reception staff in your surgery who can arrange for them to be handed to consecutive patients for completion following their appointment (rather than choosing to give them to the nice patients only!). You must continue to do this until you have at least 40 correctly completed questionnaires, which could take anything from a few days to a couple of weeks.

WHEN SHOULD I DO MY FIRST PSQ?

Your first PSQ should be done during your first general practice post in ST2. You will then need to do another in your ST3 year to make a total of two over the three years of training.

WHEN WILL I SEE THE RESULTS OF MY PSQ?

The scores and comments from your PSQ will only be released to you by your ES once at least 40 forms have been submitted and a review date has been set. Following this you can organize a formal interview with your ES for feedback, which can then be entered as a 'Professional Conversation' within your learning log.

CSR

WHAT IS A CSR?

This is a short structured report written by your clinical supervisor at the end of your hospital or ST2 GP posts. It looks particularly at the knowledge and skills gained and the competency levels attained in certain areas.

MSF

HOW DO I ORGANIZE A MSF?

This is achieved by supplying your assessors with the relevant 'ticket code' which can be accessed via the 'Evidence' link on the left toolbar of the ePortfolio and selecting 'MSF'. It is up to you which members of staff you chose to complete the assessment and hence receive feedback from. Whilst choosing colleagues that you get on well with might seem the obvious thing to do, you might be surprised by some of the comments that are made, given that they are sold as being anonymous. Requesting assessments from members of staff at random may provide you with a more accurate and valid reflection of how you are perceived clinically. It will also stand out at the panel review if all of your MSF assessments are completed by nurses or other juniors.

HOW MANY MSF ASSESSMENTS DO I NEED TO SUBMIT?

This depends on whereabouts you are in your training. In ST1 you are required to complete two MSF assessments which require responses from at least five different clinicians. In ST3 you must do the same, however you are also expected to collate responses from at least five non-clinicians within the GP surgery. Clinicians can be any member of medical staff including nurses and non-clinicians can include dispensary staff, secretaries, and receptionists.

> Tip: Requesting MSF assessments from members of staff at random (rather than using careful selection) may provide you with a more accurate and valid reflection of how you are perceived clinically.

WHEN WILL I BE ABLE TO VIEW MY MSF RESULTS?

This is only possible once all of the MSF entries have been completed and you have set a review date with your educational supervisor. Your educational supervisor can then add a comment and release the summary for you to discuss at your next review.

REFERENCES

1 RCGP. *nMRCGP update*. Available at: http://www.rcgp-curriculum.org.uk/nmrcgp.aspx (accessed 14 August 2010).

2 RCGP. *Workplace based assessment*. Available at: http://www.rcgp-curriculum.org.uk/nmrgcp/wpba.aspx (accessed 14 August 2010).

3 RCGP. *Clinical Evaluation Exercise*. Available at: http://www.rcgp-curriculum.org.uk/nmrgcp/wpba/clinical_evaluation_exercise.aspx (accessed 14 August 2010).

4 RCGP. *DOPS*. Available at: http://www.rcgp-curriculum.org.uk/nmrcgp/wpba/dops.aspx (accessed 14 August 2010).

CHAPTER 5

The ePortfolio explained

Most of you should already be familiar with the concept of an ePortfolio (having had one during your foundation years) and are therefore likely to be well rehearsed in pleading with consultants and other colleagues to complete various sections online. The GPST ePortfolio provides you with an online facility for recording evidence of your learning. Registering for your ePortfolio as early as possible is essential in ensuring that you start as you mean to go on, and don't end up leaving the task of adding your entries until the last minute. Not only does this cause stress and panic, but it is likely to be very obvious to the ARCP panel when they look through your learning log during your interim reviews (more on ARCP later…).

This chapter aims to help you get started with your ePortfolio and hopes to answer some of the more common questions and queries that others have encountered along the way. You will also find several 'help' boxes throughout the chapter, which provide you with a step-by-step approach to finding and inputting data. Whilst this may be obvious to some, there are likely to be other trainees who aren't quite as computer literate and therefore you must forgive me for the spoon-feeding!

General information

What is the ePortfolio?

The ePortfolio is a web-based learning log personal to each trainee, which is also used to gather the evidence required for the workplace-based assessment (WPBA)[1]. Once logged in you will need to use the left-hand tool bar on the home page in order to navigate your way around the various functions and areas.

How do I gain access to my ePortfolio?

First of all you will need to 'Register with the college', which can be done by following the links in Help box 5.1.

HELP 5.1

Links to follow in order to register with the RCGP online

visit www.rcgp.org.uk/register

⇩

Register with the college

⇩

RCGP self service Area

You will then be required to enter some simple details such as name, date of birth, address, phone number, and GMC number.

Once registered with the college, you will be able to apply for one of the RCGP subscriptions which can be done by following the links in Help box 5.2.

Here you will find information on the various subscriptions available and can apply by clicking on the relevant link. Ideally you should select one of the AiT subscriptions,

HELP 5.2

Links to follow in order to apply for an RCGP subscription

GP Training (top menu bar of RCGP home page)

⇩

Trainee registration
(under New Professionals)

⇩

Subscription selection

but it is still possible to gain access to the ePortfolio by choosing a non-AiT package (although this isn't recommended). The AiT packages include[2]:

- An ePortfolio
- 10% discount on AKT and CSA components of the MRCGP assessment
- Certification services (CCT only)
- Subscription to *InnovAiT*
- Subscription to the *BJGP*

Once you have completed all of the above you should be allocated a username and password to enable you access to your ePortfolio.

How do I log in to my ePortfolio?

This is done by clicking on the 'Trainee' link in the top right-hand corner of the RCGP website next to 'Access our ePortfolios'.

How do I change my password?

This cannot be done within your ePortfolio and instead must be done via the 'membership' section of the RCGP website.

How do I change my personal details?

Unfortunately this cannot be done within your ePortfolio and instead must be done via the membership section of the RCGP website. Changing your details via this route will automatically update your ePortfolio but it may take up to 48 hours to show. It is important to make sure that your personal details are correct, in particular your contact details, as they will be used by the college if they ever need to get hold of you.

When should I start using the ePortfolio?

As soon as possible! I really can't stress this point enough. It is vital that you start adding log entries from day one with a target of around two entries per week. Making the effort to adhere to this rule early on really is worthwhile, as it prevents the last minute stress that comes with 'panic logging' as review time approaches. 'Panic logging' can be very obvious to those carrying out your reviews and also won't impress those on the ARCP panel.

> Tip: Register for your ePortfolio as early as possible, to ensure that you start as you mean to go on and don't end up leaving the task of adding your entries until the last minute.

'My biggest piece of advice for the ePortfolio is to start using it early!'
ST3

How do I find out which training posts I have been allocated?

You should ideally have been given this information either in your initial offer letter from the deanery or in person by your course organizer. It should, however, also be available on your ePortfolio under the link to 'Posts' on the left-hand toolbar. It is vital that the posts information on your ePortfolio is up to date and accurate, as ARCP relies on this information when reviewing your progress.

How do I find out who my supervisors are?

This should be obvious on the 'Welcome' page of your ePortfolio. This is the default page you should be taken to every time you log in. Your supervisors can also be found under the link to 'Posts' on the left-hand toolbar. Each post should be listed, detailing the dates of that post and the name of your allocated supervisor. Your clinical supervisor should be either the consultant or the GP that you are attached to during that post, whereas the educational supervisor is normally the trainer that you have been allocated for your ST3 year. If your supervisors are not listed on the 'Welcome' page, you will need to contact the deanery to ask them to update and link this information to your ePortfolio.

Where will I find my NTN number?

This can be found within the 'Personal Details' section of your ePortfolio.

How do I sign the educational contract?

This is done by clicking on the link to 'Declarations and Agreements'. Here you will find a series of documents, including your educational contract, which you will need to read and sign online. If you click on 'Historical Declarations' you are able to view all of the previous documents that you have signed during your training.

Learning log

What is the learning log?

The learning log allows you to input educational activities of the following types:

- Clinical encounter
- Professional conversation
- Tutorials
- Reading
- Course/certificate

- Lecture/seminar
- Out of hours session
- Audit/project
- Significant event analysis
- eLearning session

The aim of the learning log is to map your learning to the RCGP curriculum and facilitate your reflection of the experiences that have arisen during your rotation. It's important to make sure that the entries you create are both relevant to your training and consistent with the competencies that you are required to attain, and not simply a list of patients that you have seen or courses that you have attended. Lists of surgical procedures that you have carried out or lectures that you have attended on the latest advances in balloon angiography are unlikely to impress anyone on the ARCP panel and you should instead focus on making your entries appropriate to general practice. There is no minimum number of entries that you are required to submit, as the focus is on quality rather than quantity and aiming to get a balance across the curriculum and competency areas.

Tip: Make sure that the learning log entries that you create are both relevant to your training and consistent with the competencies that you are required to attain.

How do I add a log entry?

Log entries are added via the 'Learning Log' link and selecting the appropriate type of log entry. This will take you to a new page where you can input the requested details and information. For each log entry you will be asked to select the curriculum headings which are appropriate to that entry. Multiple headings can be selected by holding down the 'Ctrl' button on your keyboard whilst clicking on the relevant headings. Once completed, the entry and the date it was written will be stored in your learning log.

> 'My main piece of advice would be to try and fill in your ePortfolio on a regular basis (it looks better, although it's difficult to practice what you preach!).'
>
> ST2

How do I read my previous log entries?

The learning log page lists the entries that you have already submitted in date order. However, should you wish to refine your search to display all entries of a certain type then the results can be filtered (e.g. for clinical encounter, if you click on

'Clinical Encounter' within the 'All recorded activities' link then only clinical encounter entries will be shown).

What does it mean to 'share' my log entry?

This means that it can be accessed by your educational supervisor, who on viewing the shared entry may then either add a comment or validate that entry as evidence towards your WPBA (by assigning the relevant competencies). Once this has been done, that particular entry will be 'locked', making further editing impossible. If you have chosen not to share your entry at the time of submitting it then it remains personal to you and cannot be read by anyone else. If you decide at a later date that you would like to share your entry then you can do this by clicking on the magnifying glass next to the entry and chose 'Share Record'. Un-sharing the record is done in the same way. If you chose to share your entry, a green tick will be displayed next to the date that it was written. Further green ticks will also be present if your supervisor has read or made a comment on your entry.

Can I edit the log entry once it has been submitted?

Yes. This can be done by clicking on the magnifying glass adjacent to that entry and selecting 'Edit entry'.

> 'Add things to your ePortfolio when you have the event fresh in your mind, even if it means brief notes/points, so that you can come back to it later when you have more time to increase the detail.'
>
> ST2

How do I link my log entries to a curriculum statement?

This is done by selecting the appropriate headings at the top of the input page for the chosen type of log entry. If you link your entry to a curriculum statement, then you must have the evidence within the written section and realistically should select a maximum of two to three curriculum statements, not the eight or so that the educational supervisors often see!

Also make sure that you have read the descriptors for the links, so that, for example, you don't end up matching a hospital-based clinical encounter to the practice management statement!

Can I change or add curriculum headings to an entry?

Yes. This can be done by clicking on the adjacent magnifying glass for that entry and choosing 'Select descriptors'. This enables you to add additional curriculum headings or remove existing ones which may have been added in error. You should aim to try and cover each area of the curriculum equally, in order to show that you have gained a broad range of knowledge, rather than just lots of experience in a few particular areas.

Can I see an overview of which curriculum headings I have covered within my log entries?

Yes. This can be done by selecting 'Review preparation' from the left hand toolbar on your ePortfolio home page. This not only shows you a list of your reviews but it also lists the curriculum headings and the number of the linked log entries associated with each one.

> 'Keep checking the "curriculum coverage" section on your ePortfolio, making sure that you work on filling in any gaps. You should also add the gaps to your PDP.'
>
> ST3

How do I link my log entries to the required competencies?

Unfortunately this can only be done by your ES once you have 'shared' that log entry. If on reading your entry they feel that it provides evidence of you having achieved one of the required competencies, they can then validate this by assigning the relevant competency. Over the period of three years you will be expected to have covered all of the competencies within the curriculum.

Can I upload documents to support my log entries?

Yes. If you have certificates or other types of documentation which provide further evidence of what you have achieved within your learning log, then you can do this by attaching files into your ePortfolio. This can be done by using the 'Browse' button at the bottom of the learning log entry page to upload the document from your computer. Unfortunately there is a 5Mb limit for each attachment that you upload but this should be more than sufficient. However, uploading evidence isn't essential and a quality learning log entry reflecting on what you have done and how this will benefit you in practice is just as valid and in fact usually means more than a generic course certificate.

> 'When you have completed a BMJ/CME/CPD module, try and link/upload the certificates to your ePortfolio as early as possible (otherwise you might forget later).'
>
> ST2

> 'I think the ePortfolio is a good way of providing a trainee with direction on how to focus their learning needs through reflection.'
>
> ST2

What makes a good 'learning log' entry?

A good learning log entry is concise, reflective, and should describe an activity that has made a significant addition to your training. It is also important to consider what further learning needs this activity has raised and how you will address these. Adding a long list of inappropriate e-learning modules or encounters with patients that have coughs and colds isn't really the aim of the game and instead your entries should provide good quality evidence of what you have learnt throughout your training and how you will translate it into your day-to-day practice.

> 'If you do two entries a week on the ePortfolio, that's eight a month, which is equivalent to one page. The number of months training should therefore equate to the number of pages on the ePortfolio.'
>
> ST2

PDP

What is the PDP?

Your PDP is a record of the objectives that you set yourself for your continued professional development. It should not simply be a list of all of the things that you still need to learn and you should instead prioritize the entries to those objectives which are most crucial to your training. Make sure you set yourself objectives that are both relevant to your GP training and achievable within the given time frame. PDP entries should not be entered in retrospect once you have attended a certain course or successfully completed an exam, but instead they should be carefully thought about in advance and discussed with your educational supervisor. It is not adequate to enter objectives such as 'pass AKT' into your PDP, as this is an obvious requirement for successful completion of your MRCGP.

> Tip: Make sure you set yourself objectives that are both relevant to your GP training and achievable within the given time frame.

> Tip: Consider at the start of each post what you need to learn from it and what skills relevant to general practice you might best address in this specialty. Looking at the curriculum statements might also help here

How do I create a PDP?

This is done by clicking on the link to 'PDP' on the left-hand menu toolbar. Here, you can add, edit, or view entries and confirm when each of your learning objectives has been achieved. Entries can be edited by clicking on the 'hand and tick' icon on the right side of the entry. To mark the learning objective as achieved you will need to enter the entry and tick the 'Has it been achieved' box towards the bottom of that entry. Once you have done this, the entry will no longer be 'active' but can still be accessed by clicking on 'view all' within the PDP. If you are able to upload any formal evidence of completing your learning objective, e.g. exam or course certificate, you can do this by 'browsing' the documents within your computer and attaching the relevant supporting evidence. The presence of uploaded evidence is indicated by the presence of a floppy disk symbol next to that entry. Supporting web links can also be attached. After making changes to your entry, don't forget to click 'Save' otherwise the additional information will be lost.

Entries to your PDP can also be added from your learning log once submitted. To do this you will need to click on the relevant learning log entry and then select 'Send to PDP'. This will automatically transfer what you have written within the 'What further learning needs did you identify?' section, to your PDP.

Evidence

What is the 'Evidence' page?

This shows a summary of the number of WPBA tools (e.g. MSFs, mini-CEXs, DOPS, CbDs, COTS, PSQs, and CSRs) that you have completed. By clicking on the relevant assessment tool heading you can also view a list of the submissions to date. Once you have completed the minimum number of assessments required for your next review, the number will turn from red to green. Make sure that you have selected the appropriate type of review at the top of the page; otherwise it may look like you have done more assessments than you actually have.

How do I add a COT?

This can only be added by the person assessing your COT, however a preview of the form that they are required to complete can be accessed and printed out via your ePortfolio. To do this you will need to follow the links in Help box 5.3.

This page also gives you the option to print a handout containing a 'ticket code' which your assessor can then use to input the required information within your ePortfolio. Ticket codes, however, aren't essential, as your GMC number can be used instead. To submit an entry via the main ePortfolio login page, your assessor will need to follow the links in Help box 5.4.

HELP 5.3

How to print a paper copy of the information required for a COT

Evidence (left-hand toolbar)

⇩

COT

⇩

Preview

⇩

Print-friendly version of this page (opens in new window)

HELP 5.4

How to submit an assessment onto a trainee's ePortfolio

ePortfolio login page

⇩

Assessment Forms

⇩

Complete Assessor details

⇩

Enter **'ticket code'** OR **'trainee details'** and select **'COT'**

The assessor will then be taken to an online version of the COT proforma which they can complete and submit. Once submitted the COT will show on your ePortfolio.

How do I add a CbD?

This is done in the same way as adding a COT (see earlier question **'HOW DO I ADD A COT?'**) via the ePortfolio login page, by either using a 'ticket code', or your GMC number and selecting 'CbD'.

How do I add a mini-CEX?

This is done in the same way as adding a COT (see earlier question **'HOW DO I ADD A COT?'**) via the ePortfolio login page, by using either a 'ticket code' or your GMC number and selecting 'mini-CEX'.

How do I add a DOPS?

This is done in the same way as adding a COT (see earlier question **'HOW DO I ADD A COT?'** ') via the ePortfolio login page, by using either a 'ticket code' or your GMC number and selecting 'DOPS'.

> 'Make sure that all of your WPBA assessments are logged by your supervisor on the computer otherwise all of your hard work goes to waste and isn't recognized.'
>
> ST2

How do I add a CSR?

This is done in the same way as adding a COT (see earlier question **'HOW DO I ADD A COT?'**) via the ePortfolio login page, by using either a 'ticket code' or your GMC number and selecting 'CSR'.

How do I access the required forms for my PSQ?

The forms for completion by patients can be printed out from your ePortfolio by following the links in Help box 5.5. It's probably worth printing off between 45 and 50 copies, as there are likely to be some patients that either miss sections or complete it incorrectly.

HELP 5.5

How to print off a patient satisfaction questionnaire for completion

Evidence

⇩

PSQ

⇩

Print blank form to be copied for patients to complete

How are my completed PSQs added to my ePortfolio?

Once you have completed the required amount of PSQs, you will need to organize for them to be transferred into your ePortfolio using the required 'ticket code' which can be found under the 'PSQ' section of the 'Evidence' page. The person responsible for doing this will vary depending on your deanery, but it is likely to be either your practice manager or another member of the administrative staff. Ideally all questionnaires should be entered onto your ePortfolio within two weeks of the ticket code being issued.

If, however, you fail to submit all of your 40 questionnaires within the designated time period, then you will need to gain an additional ticket code and request that the entries for all ticket codes are merged. This can be done by contacting the ePortfolio support centre via the 'ePortfolio Enquiries' link on the left hand toolbar and selecting 'Create Enquiry'. You can also contact them via phone on 0207 344 3075.

How do I organize an MSF assessment?

This is simply done by printing out the required number of ticket codes via the 'Evidence' page. Selecting 'MSF' will take you to a screen where you can view and print a 'handout' which provides the assessor with the information required to complete a form online. The number that you require and whether or not they are completed by clinical or non-clinical staff depends on the stage of your training and the post that you are completing (see **CHAPTER 4, 'WHAT LIES AHEAD? INTRODUCTION TO THE ASSESSMENT PROCESS'**). All MSF assessments need to be submitted online within two weeks of the ticket code being issued, so you may be required to do some pestering of your chosen assessors in order to get then done in time!

> Tip: Request more MSFs than you actually need. There will usually be one or more people you ask who go away on holiday or just don't get around to it and you don't want to find yourself short. They are also a really useful tool to find out more about what you need to be learning or doing to develop so use the opportunity to get feedback.

Skills log

What is the skills log?

This lists the mandatory, optional, and foundation skills which you have achieved during your training. These skills must be independently verified by way of DOPS submissions and cannot be added by simply uploading evidence in the form of documents.

In addition to submitting DOPS, you are also required to record your own level of competence for each skill and make an optional comment. This can then be edited at a later date (as you become more experienced) to change how you rate your ability in that skill. Once you have made any changes to your entry, don't forget to 'Save Comment' otherwise your changes will be lost.

Progress to certification

How do I add an out of hours (OOH) session?

This is done via the 'Learning Log' link in the same way that you would add one of the other types of learning log entry. There is no specific screen which will tell you how many hours you have completed, but instead your ES will read each of your individual entries and determine whether or not you have done sufficient sessions to sign you off as having completed this element of the assessment. You are required to complete at least 72 hours and are expected to reflect on what you learnt during each session and detail this within your learning log. It is not sufficient to simply write the location and amount of hours of each session unless your trainer has discussed and viewed the learning and certification from the OOH service and you have logged it as a Professional Conversation. It is important that the number of hours are clearly shown so that PMETB and ARCP can easily see that the required number have been achieved when it comes to your final review.

Once you have completed the required amount of 'out of hours' experience, your ES can confirm this by ticking the relevant box in your final review (it cannot be done in any reviews prior to this) which results in a green tick appearing next to this section on the 'Progress to Certification' page.

How do I get my CPR training signed off?

There is no specific area in which you should enter the details of a CPR course although the details can be submitted as a normal learning log entry. Signing you off as having completed this element can only be done by your ES during one of your reviews and requires them to tick a box confirming that you have completed the training. Once this has been done, a green tick will appear next to 'Holds valid CPR and AED certificate' within the 'Progress to Certification' page.

The deanery has sent a message to my ePortfolio. How do I check my inbox?

Your ePortfolio also acts as an additional email account, via which the deanery or college can update you on forthcoming courses or notices. It is therefore extremely important that you check your inbox on a regular basis. This can be done by clicking on 'Messages' in the top left-hand corner of your home page. Here you will be able to view the messages within your inbox and also compose messages if required. Each time a message is sent to your ePortfolio inbox you will also be informed via your regular email address, although the content of that message will not be disclosed.

Review preparation

How do I sign off a review?

Once a review has been completed by your ES you will need to 'Sign off' this review. This can be done by following the steps in Help box 5.6.

HELP 5.6

How to sign off your review

Progress to certification (left-hand menu toolbar)

⇩

Click on the magnifying glass next to **'Reviews'** (this should list all reviews to date)

⇩

Click on the appropriate review and then scroll down until you find the link which allows you to 'sign off' the review

Once signed off, a green tick will appear next to the review to indicate that it has been accepted by the trainee.

Personal library

What is the personal library?

This houses all of the documents that you have uploaded in support of either your learning log entries, DOPS, or your PDP. Additional documents which you have not linked to other areas of the ePortfolio can also be uploaded via this page should you wish to do so.

> 'The ePortfolio is a great incentive and method to make you read and learn; the only negative side of things is when a lengthy logbook entry gets lost before it's saved!'
>
> ST2

Help

Can I save the information in my ePortfolio to my personal computer?

Yes. Certain sections of the ePortfolio can be downloaded to your computer as a PDF file for safe keeping or use when applying for jobs. These are as follows:

- ARCP outcomes
- PDP
- Reviews
- Educators' notes
- Trainee details and posts

> 'Don't ignore the ePortfolio. Hit it running. It's the only way to get sufficient entries. Try to get two entries per week. I learnt the hard way.'
>
> *ST3*

Where can I go for further help?

Further written information is available via the 'FAQ' link on the left-hand toolbar or the RCGP document *The ePortfolio for GP Speciality Training*[3]. For technical information you can also try the 'Help' link or 'Create new enquiry' via the 'ePortfolio Enquiries' link. If you would rather speak to somebody over the phone you can contact ePortfolio enquiries on 0207 344 3075.

REFERENCES

1 RCGP (2008). *Associates in Training: a Guide*. RCGP, London. Also available online at: http://www.rcgp.org.uk/pdf/RCGP%20AiT%20Deanery%20Information%202008.pdf (accessed 14 August 2010).

2 RCGP. *New Professionals. Membership*. Available at: http://www.rcgp.org.uk/gp_training/aits_and_newly_qualified_gps/registration.aspx (accessed 14 August 2010).

3 RCGP. *The ePortfolio for GP Speciality Training*. Available at: http://www.rcgp-curriculum.org.uk/PDF/ePortfolio_Trainee_Manual.pdf (accessed 14 August 2010).

CHAPTER 6

Supervisors, trainers, and how to make the most of your meetings

General information

What is the educational contract?

This is a learning agreement between you and your supervisors, which along with the other declarations and agreements must be signed via your ePortfolio at the start of each post.

Who is my educational supervisor (ES)?

Your ES is the GP that is allocated to supervise your education from the start of ST1 to the end of ST3 (and beyond if you're lucky!) and should therefore remain constant throughout the entire training programme (unless you change deaneries or have a conflict of personalities!). In ST3 your ES is also likely to be your trainer.

What is the role of the ES?

Your ES is responsible for monitoring and evaluating the evidence that you contribute towards your WPBA, as well as ensuring that you receive the necessary training and experience to allow you to educationally develop at the expected rate. You are required to meet with them on a six-monthly basis in order for them to review and assess your progress towards the various elements of the MRCGP.

Who is my trainer?

Your trainer is the approved GP responsible for providing you with appropriate clinical and teaching experience during your time spent working in their surgery and in ST3 is likely to be the same person as your ES. If you spend any time in general practice prior to ST3 then your trainer for that post will also be your clinical supervisor. Confused? I thought so. Essentially, your trainer is the GP to whom you are attached during your posts in general practice, regardless of your year of training.

Who is my clinical supervisor (CS) and how are they different from my ES?

CSs are qualified specialists who are responsible for the day-to-day supervision, training, and assessment of trainees doing a placement in their specialty. Every trainee should have a named CS for each hospital placement[1] and for those in general practice during ST2, their CS will be their trainer. So whilst the ES remains constant during the three years, each CS will only be responsible for your training during the time that you spend in their specialty.

What is the role of the CS?

The CS is expected to:

- Hold review meetings with their trainee
- Use the WPBA assessment tools to record evidence of the trainee's progression towards the required competences via the ePortfolio
- Complete a clinical supervisor's report (CSR) for the trainee at the end of each placement

The formative meetings that you have with your CS at the end of each post occur in addition to the formal review meetings that you should have with your ES every six months.

How often should I meet with my CS?

You should meet your CS at the beginning, middle, and end of your placement for a review meeting. At the final review meeting your CS should complete a CSR in order to provide feedback on your performance during that post. Don't forget to log all of your meetings on your ePortfolio.

I'm struggling to get my hospital consultants on board—help!

This can be a common problem as hospital consultants are often very busy on the wards or in theatre and do not always have immediate access to a computer. Make sure that on your first meeting with your CS you schedule the three required reviews in advance rather than leaving it until the last minute. This should hopefully prevent you rushing around trying to track them down to sign off the CSR before you complete the post.

Tip: Make sure that on your first meeting with your CS you schedule the three required reviews in advance rather than leaving it until the last minute.

Your first meeting with your trainer

At the start of each post in general practice you should ideally have an initial meeting with your trainer to discuss your educational needs and any other important aspects of the job (see also Box 6.1). This is your opportunity to find out whether you have an induction timetable, when you will start seeing patients, how long your appointments will be, and what is expected of you in terms of on-call duties and home visits. It also allows you to plan your required reviews in advance and discuss with your trainer ways in which you can attempt to fill in some of the gaps in your skills log or curriculum coverage. You will also need to discuss your preference as to how you will complete your COT assessments. Some trainers are happy for you to video your consultations and then watch them back with you at a later date, whereas others would prefer to sit in on your surgeries and perform the COTs 'live' as it were.

Planning out your tutorials in advance will also help you to keep up to date with the required elements of your WPBA and prevent you from having to complete them all at the last minute as you come to the end of your post. You should be aiming to do at least

Box 6.1: A summary of questions to consider asking at your first meeting with your trainer

Summary

- Do I have an induction timetable?
- When will I start seeing patients?
- How long will my appointments be?
- When will I start doing on calls and home visits?
- How will my COTs be assessed? E.g. live or on video?
- How will I achieve my outstanding DOPS?
- When can I complete my COTs/CbDs?

one COT or CbD per month and should agree a plan with your trainer as to when these will take place.

If you still have any outstanding DOPS at the start of your post, you should make them known to your trainer so that you can benefit from any opportunities that may arise for you to get them signed off within the surgery.

Tip: Aim to do at least one COT or CbD per month and agree a plan with your trainer as to how and when these will take place.

Review preparation

How often will I have a review with my ES?

Standard full-time trainees should have a review with their ES every six months, with each review being informed by both the learning log and the evidence collected through the WPBA tools over that time period.

> 'Contact your educational supervisor early and organize an early meeting so that it doesn't get left until the last minute. It will also motivate you to get that ePortfolio looking fab!'
>
> ST2

How do I prepare for my reviews?

It is your responsibility to make sure that you are adequately prepared for your reviews, so that you can make the most of the time that you spend with your ES. There are three main areas on which you should focus prior to each review. These can all be found via the 'Review Preparation' link on the left toolbar of your ePortfolio and are as follows:

1. **Curriculum coverage** You should be able to view a list of the curriculum statement headings showing the number of linked log entries for each. Scan through the list and identify areas in which you are particularly strong and others in which you are relatively weak in terms of log entries. This should give you a rough idea of where you need to focus your future learning, which can then be discussed with your educational supervisor and an action plan made. By the end of the three years you should aim to have a roughly equal coverage of these areas in order to prove that you have the necessary knowledge and skills to become an independent GP.
2. **Skills log** Have a look through your skills log and see where you are in terms of completing your mandatory DOPS. Make sure that you update

your self-rating for each of the skills so that it corresponds with your DOPS assessments. This is also a good opportunity to highlight any procedures which you may be struggling to get signed off, as you can then discuss with your supervisor how you may rectify this over the forthcoming months.

3. **Competency list** Finally, look through the competency list and self-rate your ability in each of the 12 required competency areas. Your trainer will then complete their assessment of your competencies in each area and you can discuss any discrepancies between the two at your review. With each review you should aim for your grade in each competency area to improve, so that by the end of the three years you are achieving at least 'competent' if not 'excellent' in all 12 of the competency areas.

Once your ePortfolio is ready for review, you should inform your ES as soon as possible so that they can then spend time examining your evidence in preparation for your review meeting. On average, a review will involve between six and eight hours of their time and so it is not acceptable to announce that your review is due with only several days to go. Up to four hours will be required to read and comment on your log entries, review your PDP progress, and assess and comment upon your curriculum coverage, skills log, and competency areas. A further two-hour discussion will then be necessary in order to discuss the progress and agree on a plan for any outstanding learning needs.

> 'Regularly check your curriculum coverage as there are things that you just won't come across that you will need to find an alternative way to cover.'
>
> ST2

How long should each meeting last?

This will depend on how much material you have to cover and how well prepared you were for the meeting. One to two hours should allow sufficient time to discuss the necessary elements of the review but it is your responsibility to make sure that you schedule in enough time for these sessions in advance of your ARCP review date.

What happens at the end of the review?

After completing each review your trainer will release the information for you to view on your ePortfolio. You must then agree to what is written by 'signing off' the review. A completed review that is awaiting your approval should be obvious on your ePortfolio homepage by stating that you have '1 unsigned Educational Supervisor's Review'. Clicking on this link allows you to view the review and accept it by clicking on the 'Trainee Signoff' link. Further information on how to do this can be found in **CHAPTER 5, 'THE EPORTFOLIO EXPLAINED'**.

Once your review has been signed off, all WPBA tools completed during that period will be locked and cannot be carried forwards to subsequent reviews.

Whose responsibility is it to organize the reviews?

Yours I'm afraid. Make sure that you give your supervisors sufficient warning as to when your reviews are due and allow them to schedule your meeting into their timetable. They are likely to be very busy people who cannot squeeze in a quick meeting at the last minute and instead will need a certain amount of time to read through your ePortfolio and do the necessary preparation in advance. It is not reasonable to have updated your learning log at midnight and expect the ES to be prepared for a 9am review!

When do the reviews need to be done by?

You are required to have a review with your ES on average every six months, however the second review of each year is normally done slightly earlier so that it is completed in time for your ARCP review. This means that you have more like four to five months in which to complete the required number of WPBA assessment tools in the latter half of each year, which is only a problem if you are not prepared for it.

What if I'm training flexibly?

Flexible trainees will still be required to have a review every six months and must complete the same minimum number of DOPS, mini-CEXs, CbDs, and COTs as full-time trainees within this time. MSF and PSQ, however, should be presented at the review that represents the mid- and end-point for ST1 and ST3 respectively. This ensures that there is sufficient time for appropriate feedback and any necessary change in behaviour. More detailed information on this can be found on the RCGP website page 'MSF and PSQ for less than full-time trainees (LTFTTs)'[2].

REFERENCES

1 RCGP. *Clinical Supervisors*. Available at: http://www.rcgp-curriculum.org.uk/gp_training_information/clinical_supervisors.aspx (accessed 14 August 2010).

2 RCGP. *MSF and PSQ for less than full-time trainees (LTFTTs)*. Available at: http://www.rcgp-curriculum.org.uk/nmrcgp/less_than_full_time_trainees/msf__psq.aspx#LTFTT_evidence_schedule (accessed 14 August 2010).

CHAPTER 7

The money stuff

When you started thinking about a career in medicine you were probably under the illusion that once qualified you would become incredibly rich, buy yourself a totally impractical fast car, and frequent the Seychelles for your holidays. If only back then you had realized that in fact once you started your training you would be handing over your credit card details more often than you had hot dinners, then you might have had second thoughts. But then of course, we all know that no one really goes in to medicine for the money. We do it because we want to help people and enjoy solving the mysteries that are our diagnostic dilemmas (or at least that's what we told the panel at our medical school interviews!).

Not many GPs these days actually earn the six-figure salaries that are so often moaned about in the tabloids, although there are obviously some partners working in high-earning practices that do. However, they are certainly not in the majority and with fewer practices taking on partners and more leaning towards recruiting extra salaried doctors, the pay that you will eventually receive may not seem as attractive as it once did. GP registrar (ST3) pay is also on the decrease and has been for the last few years, so it's worth bearing all these factors in mind and starting to think about budgeting for the training years ahead. The next three years will end up costing you a considerable amount of money, even if you only fork out for the basics and don't splash out on any fancy books and courses. However, it's not all doom and gloom, doctors in general are still paid reasonably well when compared with other vocations, and they have the benefit of the NHS pension and job security, so it's worth remembering these things when you start to feel hard done by. It's also worth noting that your colleagues training in the hospital specialities will also be forking out the same sort of sums of cash—so don't let the money stuff put you off general practice.

This chapter helps you to prepare in advance for the expenses that you will definitely encounter, plus lets you in on some of the hidden ones that may not seem obvious at the start. Sensible budgeting will make a huge difference in the long run and means that you can also try and factor in a well-deserved holiday!

The outgoings*

ST1

RCGP REGISTRATION

This ranges from £142 to £532 depending on your ST level. For further information see **CHAPTER 3** question **'HOW DO I REGISTER WITH THE RCGP?'**

BOOKS

As discussed in **CHAPTER 3**, the most useful book that I would advise you to buy in ST1 is the *Oxford Handbook of General Practice*. The 3rd edition was published in 2010 and was retailing at £32.95 in the RCGP bookstore (although as an AiT you will receive 10% discount off this price).

You may also want to consider buying your own copy of the *BNF*. Whilst most hospitals should have an accessible copy on the ward, you may not get the up-to-date version given to you in primary care and having your own copy means that you can highlight important sections and put markers on pages that you frequently use. It can also come in handy when you start revising for your exams, as some of the AKT questions are based on information that can be found in the *BNF*. At the beginning of each chapter there is a useful section on that specific group of drugs, including brief information about the conditions they can be used for and up-to-date guidelines on how they should be prescribed.

DIPLOMAS

You may decide whilst doing one of your hospital posts that you would like to consolidate and formalize your knowledge and experience by working towards the diploma in that speciality. There are many diplomas available for you to achieve if you so wish, however they don't come cheap and do require a considerable amount of work if you want to do well. Further information about the different diplomas available and how to go about organizing them can be found in **CHAPTER 13, 'THINKING AHEAD AND HOW TO MAKE YOURSELF EMPLOYABLE AT THE END OF IT ALL'.** At this stage, you will need to consider which extra qualifications you would find both useful and enjoyable and consider how much they will cost you. By the time you have paid to sit the actual exam (or attend the relevant training), subscribed to any online revision resources, or bought any study aid books, you will probably end up spending at least £500 for the pleasure of having those added letters after your name. Some diplomas will cost you considerably more than this and some may be less, but whatever the cost, you need to make sure that you budget accordingly.

*Fees and other costs are accurate as of 2010, but may since have changed.

ST2

MEMBERSHIP EXAMS

AKT

Whilst the RCGP recommend that you sit your exams in ST3, some of you may decide that you want to attempt the AKT in ST2. If that's the case then you will need to find around £400 for the pleasure. If you are unsuccessful you will be required to pay the full amount again as there are no discounts for second or third attempts!

ST3

MEMBERSHIP EXAMS

AKT

See previous heading under **ST2.**

CSA

Sitting the CSA will cost you around £1,400 for each attempt (although those who joined the RCGP as associates in training will receive 10% discount off that price taking it to £1,260)[1]. No one wants to imagine that they will ever fail the CSA, but unfortunately a small proportion of people do and are therefore required to fork out the one and a half grand (there or thereabouts) for every attempt that they have. As with AKT, there is no 'buy one get one free' and you do not get any discount for subsequent attempts. It is therefore essential that within your budget you consider the fact that you might not pass first time and will thus need an extra pot of money to allow for re-sitting at the next session. Finding this sort of money at short notice isn't easy for anyone and you don't need any extra stress when having to re-sit such an important exam. You should also factor in the cost of your travel and accommodation expenses, as whilst you may be able to claim some of this back from the SHA, you will need to pay for this upfront and it may be some time before you are reimbursed (if at all!).

CCT

Once you have done all of the hard work and are signed off by the ARCP panel as having successfully completed your MRCGP, you will then need to apply to PMETB for certification. The 2009/10 fee for this was £805 which needs to be paid in full at the time of application.

DOCTOR'S BAG

Towards the end of your ST3 year, you may also want to consider buying yourself a doctor's bag and some of the more useful things that you may require. You are likely to already have your own stethoscope, but other things which you may want to consider buying are an otoscope, ophthalmoscope, digital thermometer, peak flow monitor, or even an oxygen saturation monitor if you're really wealthy! These items can be expensive and if you are planning on doing locum work (or even if you have a salaried job), you are likely to be required to provide your own equipment so will need to buy them at some point. Purchasing a specific bag designed for carrying medical equipment isn't essential (and they can be quite pricey!) as many people use a simple holdall or laptop bag to the same effect, but at a much lower price. Remember that BMA members get discount on medical supplies when you buy through them so bear this in mind when considering where to shop.

Annual expenses

RCGP ANNUAL SUBSCRIPTION (AIT)

This is completely separate to the initial registration fee and is the amount you are required to pay on a yearly basis in order to continue your membership with the RCGP as an AiT.

GMC

The annual fee for full registration in 2010 was £410 which can either be paid as one amount or by direct debit in quarterly or 10 monthly instalments. You are not charged any extra by paying in instalments and it might just help when trying to budget for your expenses. Make sure that you if you do want to pay by direct debit that you complete the necessary forms well in advance of your renewal date, otherwise you will miss the boat and be expected to pay the full amount in one payment until your direct debit kicks in for the following year.

If your annual income is less than £21,862 then you are eligible for a discounted rate on your GMC subscription, although this is unlikely to apply to many of you, if any.

If you don't pay your annual retention fee on time, then your name may be erased from the register and your licence to practice revoked. Hence, the direct debit option is probably the safest—it's an important payment not to miss!!

MEDICAL DEFENCE ORGANIZATION MEMBERSHIP

You should already have membership with one of the defence organizations, but remember that prior to starting any posting within general practice you will need to update your medical defence organization. This is because changes in your working circumstances may have an impact on your subscription and it is important to ensure you are sufficiently indemnified for the work you are undertaking. It's worth bearing this in mind in the months running up to starting the post, as whilst most medical defence organizations will let you pay for your indemnity by installments to spread the financial burden, you will still need to find the cash out of your own pocket until you get reimbursed from your practice (via the PCT). Some practices will reimburse you monthly in your payslip whereas others may give you a cheque as and when they receive the money from the PCT. You should also be aware that you will only be reimbursed for the extra cost of the general practice component and will still be required to pay the remaining amount that you would have paid for your hospital jobs.

BMA

If you decide to become a BMA member or plan to continue your BMA membership, then the standard annual rate for 2010 was £407. If, however, you have been qualified less than seven years then you are eligible for a reduced rate, which ranges from £104 to £358 per year depending on how many years post qualification you are. Current rates can be found online in the section 'Join BMA' on the BMA website[2]. Interestingly, however, if you have been a member for more than 50 years then you are entitled to free membership (although I'm guessing this won't apply to any of you!). For those of you who are outside of the seven years post qualification and are either in the armed forces or a resident of the Channel Islands or Isle of Man, for some reason you also get a reduced annual membership rate of £358. Married couples or partners in long-term relationships are also able to get a reduced rate of membership, although the reduction is only applied to the partner who would have paid the lesser rate. The downside is that you'll have to share your copy of the *BMJ* with your loved one although you will halve the load of recycling that you'll need to take to the tip on a Sunday. How you prove that you and your 'partner' are in a long-term relationship I'm not sure, although you do have to live in the same home. Given that medics often end up in relationships with medics (we certainly had loads of couples on our VTS scheme!) it's worth knowing these sorts of things in order to save yourself money wherever you can!

GENERAL EXPENSES REQUIRED FOR GP TRAINING

- Running a car—make sure that your tax and insurance are up to date and that you inform your insurance company that you will be using your car for business purposes during your GP posts. This shouldn't cost you any extra but is essential, as if you are involved in an accident whilst on a home visit and your insurance company are not aware that you require your car for business use, they may refuse to pay out. Also check that your policy entitles you to a 'replacement vehicle' should you have the misfortune of writing off your car, as without this you will be left with the stress of how you're going to get to work.

> Tip: Check that your car insurance policy entitles you to a 'replacement vehicle' should you have the misfortune of writing off your car, as without this you will be left with the stress of how you're going to get to work.

- Owning a mobile phone—preferably one that has both credit (if pay as you go) and decent signal, because if your phone constantly goes to answer phone every time someone tries to phone you whilst you're on call, people will get frustrated and it doesn't give a particularly good impression.
- Going on holiday—very important! No one can survive without having some time off and a decent break. Never underestimate this and make sure that you budget for this on an annual basis if possible.

The incomings

Salary

The amount of salary you receive will depend on the post that you are doing and how many years post qualification you are. As with your previous hospital posts, you will receive a basic salary plus a supplement in accordance with your post banding. Posts with no unsociable hours are very unlikely to attract any additional pay and you may therefore receive basic pay only. This is discussed further in **CHAPTER 1** question **'HOW MUCH WILL I GET PAID?'** and **CHAPTER 3** question **'DOES THE ROTATION THAT I DO AFFECT MY PAY?'**. Those entering a training placement after April 2009 may only get a 45% supplement for the time that they spend working in general practice, and in subsequent years it is likely to become even less.

Claiming money back (e.g. medical defence organization expenses, study budget, and car allowance)

There are several ways in which you can claim money back for some of your outgoings. Additional medical defence organization membership subscription fees for your general practice posts can be claimed back through your surgery on production of a valid certificate proving your payment. Money spent on courses may possibly be recovered from your study budget if there are sufficient funds available and it is deemed relevant and useful to your general practice training. You should be able to claim for travel and subsistence expenses for postgraduate exams although the actual exam fee cannot be claimed[3]. Claim forms (GPCF3) may be obtained from the Strategic Health Authority, and should be returned to them after your GP trainer has completed the form to certify that you attended the exam.

The GP registrar car allowance was abolished some years ago but you are still able to claim for the mileage that you do whilst in a GP post. You are only allowed to claim for your commute to work if you were required to do a home visit that day and can obviously claim for your home visit mileage in addition to this. Depending on how much mileage you are required to do, you can either apply for a small lump sum payment or claim for each mile that you travel. The majority of people are only eligible to claim per mile, so make sure that you keep a weekly record of your mileage whilst in general practice so that you aren't desperately trying to guess when it comes to the end of the post. The practice manager at your surgery should be able to give you more information on how and what you can claim back.

> Tip: Make sure that you keep a weekly record of your mileage whilst in general practice so that you aren't desperately trying to guess when it comes to the end of the post.

Tax rebates

Many people find the whole tax rebate situation very confusing and are unsure what can and can't be claimed back and how you go about it. If you're keen to make sure that you are doing everything by the book, then for about £200 a year you can enlist the help of an accountant who can sort it all out for you. Remember that if you are trying to claim money back from the tax man then you must make sure that you have been honest about any extra income that you may have earned in addition to your salary during that tax year (including cremation fees), otherwise you could find yourself in a fairly expensive battle with the Inland Revenue. Expenses that you can claim tax back for include AiT registration and subscription but not exam fees such as AKT and CSA as these are 'deemed *preparation* for performing the duties of an employment rather than incurred in the *actual* performance of the employee's duties'[4].

Cremation forms

You will no doubt have had experience of completing cremation forms and receiving payment for these during your hospital posts and therefore this may not be a new concept. However, whilst working in general practice you are unlikely to receive any money for completing these as it will normally be the practice that will receive payment rather than you personally. Some surgeries may have different policies to this, however this seems to be the general rule.

Locum work

With the drop in banding supplement during your GPST posts plus the lack of supplement offered by some of the new posts such as public health or palliative care, you may be tempted to carry out some locum work during your training years in order to boost your income. Some trainees sign up for work in their local A&E department during their spare time, in the hope that the extra money will help pay the mortgage and perhaps afford them a holiday. Whilst this may sound appealing, think carefully about what you commit yourself to. Studying for your membership exams is hard work and it would be a mistake to jeopardize your success in your training just to make a few extra pounds. You may also find that the deanery refuse to extend your training should you fail your exams, if they find out that you were working in addition to your GP training and therefore didn't optimize your chances of passing.

Any work that you do sign up to outside of your training programme should be discussed with your trainer, in order to not only make sure that you are complying with European Working Time Directive (EWTD) but also to prevent you from 'burning out'. The BMA contract states that with the agreement of your trainer/educational supervisor, 'you may arrange to undertake additional duties or professional activities outside those of the practice however they must not compete with or impinge on your normal contracted duties or your GP vocational training. This applies equally whether such duties are remunerated or not. You are also advised to ensure that your membership of a recognized medical defence organisation is commensurate with these activities'[5].

Gifts

From time to time you may receive gifts from patients, which are usually in the form of cards, chocolates, flowers, or wine. Small gifts such as these are acceptable for you to keep without having to worry about declaring them to anyone, although larger gifts including money should ideally be discussed with your trainer, as monetary gifts of over £100 need to be declared to the PCT.

Prizes and awards

Every year the RCGP have several opportunities for trainees to submit work in order to win prizes or awards. These are often also available at a local level and do not seem to be very well publicized. As a result, those that do enter have a reasonably good chance of winning. Prizes often have monetary value and therefore putting in a bit of extra work to make your audit project look impressive enough to submit, will not only look good on your ePortfolio but might also help to pay off your credit card bill. Check out the section entitled 'Prizes and Awards'[6] on the RCGP website.

Summary

Hopefully reading this chapter has highlighted the various costs involved in GP training and has provided you with an insight into what you need to budget for at the beginning of your three (or possibly five) year rotation. For a summary of the costs that I encountered during my training see Table 7.1. Prices have obviously changed over the last few years so I have tried to adapt it accordingly.

Table 7.1 The cost of my GP training (adapted based on 2010 equivalents)

Activity	Year	Average cost in pounds (£)
RCGP registration	ST1	£142
AiT annual fee	ST1	£213
GMC annual fee	ST1	£410
MPS Hospital	ST1	£80
BMA Year 3	ST1	£205
Oxford Handbook of General Practice	ST1	£30
DRCOG online revision (4 months)	ST1	£90

DRCOG exam	ST1	£375
AiT annual fee	ST2	£213
GMC annual fee	ST2	£410
BMA Year 4	ST2	£205
MPS GP/Hospital	ST2	£80–750 (average £330)
DFSRH theory course	ST2	£400
DFSRH practical	ST2	£400
Minor surgery course	ST2	£450
AKT	ST2	£390
AKT revision books	ST2	£30
AKT online revision (4-month subscription)	ST2	£80
AiT annual fee	ST3	£213
GMC annual fee	ST3	£410
BMA Year 5	ST3	£303
MPS GP Registrar	ST3	£1,600
Doctor's bag and equipment	ST3	£300
CSA	ST3	£1,400
CSA books	ST3	£50
CSA course	ST3	£400
CCT	ST3	£805
TOTAL	**ST1–3**	**£9,934**

Whilst you will be able to claim back the medical defence organization subscription and may get some study budget refunds for courses attended, there are likely to be several other expenses to consider that have been missed from this list. Based on these figures, however, the cost of GP training appears to be on average around £3,000 per year. This of course takes into account doing extra diplomas and courses but even if you only paid for the essential subscriptions and exams then you would still need to find around £5,000 over the three-year period. This table, however, doesn't include reserve funds for having second or third attempts at exams which can add considerably to the expenditure. Essentially, taking this into account you need to consider putting aside around £200 per month over the three years in order to allow for the above costs (this does include a re-sit fund which can be used to pay for a holiday if you are lucky enough to pass first time!).

REFERENCES

1 RCGP. *Frequently asked questions*. Available at: http://www.rcgp.org.uk/gp_training/aits_and_newly_qualified_gps/registration/faqs/csa.aspx (accessed 14 August 2010).

2 BMA. *Join BMA*. Available at: http://www.bma.org.uk/_top/join_bma/subscriptionrates.jsp (accessed 14 August 2010).

3 NHS employers. *Schedules to Direction to Strategic Health Authorities Concerning GP Registrars (2003) with 2009 Amendments*. Available at: http://www.nhsemployers.org/PayAndContracts/JuniorDoctorsDentistsGPReg/Pages/DoctorsInTraining-GPRegistrars2007.aspx (accessed 14 August 2010).

4 RCGP. *Your fees and allowable expenses*. Available at: http://www.rcgp.org.uk/gp_training/aits_and_newly_qualified_gps/registration/faqs/allowable_expenses.aspx (accessed 14 August 2010).

5 BMA. *Framework for a written contract of employment for GP specialty registrars*. Available at: http://www.bma.org.uk/employmentandcontracts/employmentcontracts/junior_doctors/framecontractGPregs0707.jsp (accessed 18 January 2010).

6 RCGP. *Prizes and awards*. Available at: http://www.rcgp.org.uk/news_and_events/prizes_and_awards.aspx (accessed 14 August 2010).

CHAPTER 8

Your hospital posts and how they differ to the foundation years

As part of your ST1 and ST2 years you will be allocated posts within the hospital setting in order to gain additional experience in more specialized fields of medicine. The posts that you are allocated will vary massively depending on when and where you are training, but are likely to include some of the following:

- A&E
- ENT
- O&G
- Palliative care
- Paediatrics
- Psychiatry
- Acute medicine
- Elderly care
- Public health

Whilst within some deaneries there may be scope for swapping your allocated hospital posts with another trainee (should you be allocated a speciality in which you have already worked) this is likely to become more and more difficult as the number of trainees increase, is probably more hassle than it's worth, and is certainly not possible within most areas.

No trainee will have the opportunity to work within every speciality, but instead they will be offered a rotation which provides variety and, more importantly, complies with that required by the PMETB for certification. Swapping posts between trainees would therefore lead to problems, as you may inadvertently find yourself doing a combination of posts which are not compatible with gaining your CCT. Trainees often complain that not being given a post in paediatrics or obstetrics means that they will struggle to achieve the knowledge and competencies required within the curriculum for that field, however these shortfalls can often be accommodated for during your general practice posts by way of attending additional clinics, lectures, or courses.

Working as a GP trainee within a hospital post can sometimes be difficult, as you are no longer under the umbrella of being a foundation doctor and despite being more senior in terms of years post qualification, may have little or no experience of the field within which you are working. But remember, you have spent the last few years training to be a doctor and as long as you apply the basic principles to your clinical work and most importantly recognize your limitations, then you can't go too far wrong. However, in order to help you along the way, I have created my top 12 hints and tips to help you succeed in your hospital posts.

1. **Check the RCGP curriculum** At the start of each hospital post, read through the corresponding section of the RCGP curriculum and set yourself objectives detailing what you would like to achieve during that post. You are obviously not going to become an ENT specialist in six months, but what you can try to do is obtain the relevant practical skills and clinical knowledge that are detailed in the curriculum for that speciality.
2. **Read the relevant chapter in the *Oxford Handbook of General Practice* (or a similar book) before you start** This will make the first few weeks of your job much easier, as you will have refreshed your knowledge on some of the more common presentations and conditions that you are likely to encounter within that speciality. Some of you may have only ever read about O&G for example, and will have never actually worked on an O&G ward or in a clinic. Revising some of the basics beforehand therefore is likely to make you approach the job with more confidence and prevent you from looking totally clueless on the ward round on your first day.
3. **Make friends with the nurses on your ward** Never underestimate how helpful the nurses on the ward can be when you first commence your post. You will probably already have realized the importance of this during your foundation years. Many of them have been working in that field (and sometimes even that ward!) for years and therefore have a wealth of experience and knowledge from which you can learn. They will also know the practical stuff such as where the venflons are kept and how you organize an outpatient appointment. Respecting the nurses and other members of staff will make the wards a much more pleasant place to be and if they like you, they might even make you a cup of tea when you're rushed off your feet on call! (But never expect them to—that's one way of getting off on the wrong foot right from the start!)

4. **Try not to be intimidated by the hospital trainees** Working as a GP trainee on the same rota as the hospital trainees can sometimes be daunting, as they have obviously been doing the job a lot longer than you and are likely to have more skill and experience. However, try not to let this intimidate you and instead take every opportunity to learn from them what you can.
5. **Try to attend the outpatient clinics** Sometimes in busy hospital jobs there is not the time for the junior doctors to attend or help out in the out-patient clinics, as the ward work and on-call duties often take priority. However, if you are given the opportunity to attend clinics then I urge you to do so. Very often you will be seeing common problems that could have probably been managed in the community if the referring GP had known the local policies and preferred management options. Going to clinic and acquiring this 'insider's knowledge' will be extremely helpful to both you and the rest of the team at your surgery and might enable you to defer a referral of a similar nature in the future.

   'Try to always attend clinics and observe skills that will better your understanding and help to improve your knowledge as a GP, e.g. postmenopausal bleeding clinics in your O&G post.'

   ST2

6. **Make the most of every opportunity presented** When you know that at the end of it all you want to be a GP, it can be tempting for some to coast through their hospital posts trying desperately to avoid spending hours in theatre or getting interrogated on ward rounds. However, the hospital posts are largely pretty good fun, so instead try to make the most of the opportunities presented to you within each speciality, as you are probably never going to get this chance again. If you have the time to go to theatre and think that it will be of benefit to you then do it, but not at the expense of perhaps a more useful outpatient clinic or teaching session. Knowing what happens in operations, however, can be very useful, as patients will often come to you in general practice to discuss certain procedures that they are due to have, as they perhaps perceive you to have more time than the hospital consultants to chat through things. You probably don't need to see each procedure more than a couple of times though and I would suggest that there are probably better ways of spending your time than hours of 'scrubbing up' or holding a retractor.

   'In the hospital jobs you may struggle to get your assessments and forms done—try and use every opportunity from day one of the post and make sure you tell your supervisor what is required at your first meeting.'

   ST2

7. **Teach your juniors** Now that your foundation years are behind you, you are no longer the baby of the team and instead will be looked up to by the more junior colleagues and medical students attached to your firm (well... maybe!). Take the opportunity to get involved with some teaching as not only will it look good on your CV but as a wise person once said to me 'the best way to learn is to teach'. You are likely to have more spare time than the registrars and consultants working on your ward and are also more likely to be up to date with the medical student curriculum and foundation year's assessment process.
8. **Learn from your seniors** Having the opportunity to work alongside a specialist for six months is extremely valuable, as not only will you become aware of the local policies and procedures within the hospital, but you will also learn firsthand how to clinically assess and treat conditions that you may have only read about in books. Remember to stay focused during your learning though and try not to get lost in some of the weird and wonderful rare cases that you might see. Instead make sure that you feel competent in managing common problems that present frequently within that speciality (and are therefore likely to be things that you will see and be able to manage in primary care). It's also a fantastic opportunity to polish your examination and practical skills, as there is more opportunity for a senior doctor to observe, assess, and correct your technique within the hospital setting, which can be invaluable to your further medical education (and means that you can sign off plenty of DOPS!).
9. **Write up interesting cases** Writing up interesting cases or making brief notes on what you are learning on a day-to-day basis can be extremely helpful when you are faced with a similar patient in general practice and have forgotten what to do. Within your hospital posts you will often be seeing the same sorts of cases on a regular basis and will therefore be able to manage these patients with your eyes closed whilst you are working there. However, once back in primary care, where you will probably only see that condition once in a while, you may appreciate those brief notes you made during your hospital post as they will jog your memory as to how that condition is managed in your local area. Formally writing up interesting cases via your ePortfolio means that you will also keep your learning log up to date and have a permanent record of what you have been learning during your post.
10. **Try to do an audit** (but make it relevant to general practice if you can!) This will not only be a good addition to your ePortfolio but might enable you to improve either your own clinical practice or that of your surgery in the future.
11. **Learn some specialized skills** This one isn't essential but if you get the opportunity to learn some of the additional skills listed in the skills log then it won't do any harm. Don't bust a gut to get these done

though if there isn't an obvious opportunity. They might just come in handy once you're out in general practice if you decide to get involved in some of the additional services offered in your surgery.

12. **Prepare for your move back into a general practice at the end of your posts** There are a few very important things that you will need to do before the end of your hospital posts and the start of your general practice posts. The two most important of these are as follows:
    - *Change your medical defence organization status from a hospital trainee to a GP registrar*
    - *Check that you are included on your local Performer's List (and if not submit your application as soon as possible!)*

For more information on preparing for the move back into general practice see **CHAPTER 11, 'YOUR FIRST GPST POST AND SOME BASIC TIPS ON HOW TO SURVIVE'.**

CHAPTER 9

How to get the best out of your regional VTS teaching

Within each deanery and its subregions, there are likely to be vast differences in the way in which the VTS teaching and training is delivered. Some schemes have a set time each week where all of the trainees meet for teaching, whereas others will have a series of full days or blocks of days during which they are expected to attend. Finding out what is on offer within your scheme is essential and is often possible via the individual websites (if they have one!). The VTS teaching sessions not only provide you with the opportunity to meet up with your peers and compare training experiences, but they also give you the chance to learn some of the more complex topics detailed in the curriculum.

Many trainees starting out on their VTS programme are keen for the teaching sessions to provide up-to-date knowledge on some of the more specialized areas of medicine such as ENT, dermatology, and ophthalmology, as these are often areas which are taught badly at an undergraduate level and are not encountered on a regular basis during general hospital rotations. However, whilst in the past, many schemes may have tried to get expert speakers on these subjects in an attempt to rectify this gap in most trainees' knowledge, it is not necessarily the best use of the teaching time available when there is so much else within the GP curriculum to cover. The general thinking is that much of the specialist knowledge in these areas can be gained by reading the relevant books and learning from experience rather than a whole host of long (and often far too specialized) lectures. You have to remember that consultants tend to be fascinated by their own speciality and therefore sometimes fail to recognize that whilst you may be impressed by their latest technique of resurfacing a hip, this will be of little value to you in your general practice career.

Instead, your TPDs are more likely to focus on the less clinical aspects of the curriculum, such as practice management and ethics. Some schemes may ask their trainees to indicate at the start of each year which topics they would like to focus on during their teaching sessions. If this is the case, think carefully about what would benefit you most and more importantly, let them know. It makes it very difficult for you to complain about the quality or content of your teaching if you are not providing feedback or alternative suggestions. You are very unlikely to be spoon-fed in terms of your education within general practice training and therefore your contribution is essential in order to achieve the necessary goals set out within the curriculum. In fact, the most valuable sessions (in which the deepest learning takes place) are invariably those run by the trainees themselves. If you are given the opportunity to put on a session or part of a session, volunteer—it's a sure way of finding out what you need to know when you try to help someone else to learn (plus you can add to the teaching section of the ePortfolio!).

A good scheme will usually deliver the education in different ways, get everyone involved, and hope to get the balance right most of the time. Prepare to participate, make the most of the opportunity, and you'll enjoy every minute!

> 'Talk, talk, talk to fellow doctors about your patients to share your experiences—it's a great learning experience!'
>
> ST2

The following questions are some which are often asked in the early days of commencing the training programme, the answers to which should hopefully be of some assistance to you.

How often will we have teaching?

This varies between each VTS scheme from either one afternoon a week, one full day every couple of months, or a residential cluster of days twice a year. Essentially, anything is possible but there should definitely be formal teaching of some sort.

Will the ST1s, ST2s, and ST3s all have teaching together?

In some areas yes, but in others they are split depending on year of training or into small groups containing trainees at different stages.

Where will the teaching be held?

It is likely to be held at one of your local hospitals or perhaps a central conference centre or hotel that is fairly accessible to all trainees.

What will the teaching involve?

Again, this will vary hugely between each scheme but remember that the final aim is to assist you in passing your MRCGP and becoming a good GP. It is therefore likely to be based around the specific topics within the RCGP curriculum. The style of VTS teaching, however, doesn't suit everyone's learning style and therefore don't be too disheartened if this applies to you—you won't be the first person to go away thinking 'what on earth was all that about?', particularly at the start of your training.

> Tip: Don't be too disheartened if you go away from the first few teaching sessions thinking 'what was all that about?'—you won't be the first person to whom the teaching style doesn't match their learning style, but it'll get easier after the first few months.

Will there be any 'clinical' teaching?

Possibly, although most VTS schemes are steering away from this in favour of covering some of the more non-clinical areas of the curriculum.

> 'Grab every opportunity for clinical teaching—the sessions may not be in your practice, team, or department but if you don't grab what you can then might regret it at a later date!'
>
> ST2

All we seem to do is play games—is it just a waste of time?

Believe it or not, many of these games are organized in order to aid team building and develop trust within the group. Some, however, are plain ridiculous and provide no educational benefit whatsoever (but instead prove to be incredibly amusing and help to lighten the mood!).

Will our attendance be monitored?

Yes. As part of your commitment to the scheme you will be expected to attend a certain number of teaching sessions, with special allowances made for sick, annual, or maternity leave. If your attendance falls below a certain level you are likely to find yourself having a difficult conversation with your trainer, if not someone within the deanery. Essentially you are still being paid for the hours during which you are expected to be at the teaching, therefore it really isn't acceptable to see this time as extra holiday and choose not to attend.

Tip: If you have been to a particularly useful teaching session don't forget to add it to your ePortfolio with a description of how it will affect your future practice!

My consultant won't release me from the ward for teaching—what can I do?

This is a difficult one but unfortunately a very common problem. Whilst your education is extremely important and should ideally be factored into your working timetable, in some specialities it may simply not be feasible given the level of service commitment required in that post. If this is happening on a regular basis then you should discuss it with your clinical supervisor in the first instance and then your educational supervisor if you have no success. Most training rotations have dedicated time in which you should in theory be 'bleep free' with someone else covering your ward or on-call duties, but this isn't always possible. Make sure that you attend as many of your teaching or study days as physically possible, because if you don't and have not given valid reasons for this, then you could find yourself having to explain why to someone at the deanery.

What are the 'GPST clusters' that I've heard about?

Well, they aren't a new variant of breakfast cereal and are, in fact, at present a sort of 'work in progress'. It may well be that over the next 12 months, large group VTS training will be replaced by small group 'cluster' teaching based in GP surgeries to allow more concentrated teaching.

PART 3

Early ST2

CHAPTER 10

The move into general practice and how it differs from hospital medicine

I have been there too. Receiving your fifth call of the night from the GP in out of hours when you're on call for medicine at the hospital and have only been there an hour, can be totally draining and exasperating. It's similar to that feeling you get when you open the fiftieth GP letter of the day to find that not only is it illegible with no medication list attached but also has a coffee mug stain at the top. The frustration caused by the lack of information provided on the patient doesn't even come close to the annoyance that builds up when you imagine the GP in question sipping his or her fifth coffee of the day, whilst you, however, are pounding the hospital corridors having not eaten or drank for 12 hours. But when you have only ever worked in hospital medicine and have never experienced a day in general practice you really don't have any idea what pressures most GPs are under on a daily basis. It's not until you've spent perhaps a few weeks or months in the job that you understand how much more there is to general practice than just drinking coffee and sending patients in to hospital for a second opinion.

Fundamentally there are huge differences in the way in which medicine is practised in the community when compared to the hospitals. This chapter aims to open your eyes to some of these differences, in the hope that not only will it make you better prepared for general practice but it will also stop you from becoming one of those irritating junior hospital doctors that you get on the phone, who wants to know whether or not you have done a pregnancy test on a 60-year-old lady that you are trying to admit with acute abdominal pain.

The following issues are things that you may not have thought about and are worth considering before your transition into general practice.

Uncertainty

This is probably the biggest hurdle that you have to face in general practice and may be a concept that until now you have never really had to consider. When you see patients in the community you can talk to them, examine them, maybe even do a few basic tests but you can't put them in a bed in the corner of your room to 'observe' them as the day goes on and you certainly can't just take some blood and order a chest x-ray with a view to making a decision in a few hours' time. Of course, you can in theory order tests (although you would be unlikely to get any results back that same day) and you could in theory put them on a couch in another room in order 'to wait and see', but you can't do that with everyone, as there is neither the time nor the space. Instead you have to make a decision based on your findings there and then and deal with the level of uncertainty that goes with it. In some ways, therefore, you are making a hypothesis rather than a diagnosis as such and will need to act on this hypothesis accordingly. For some doctors this 'uncertainty' is too much to cope with and the nights spent lying awake at night wondering whether Mrs Jones's indigestion was in fact a heart attack, prove too much. Of course, this fear of uncertainty highlights the importance of what we refer to in the trade as 'safety netting', but that will be discussed in more detail later. You must remember that you are by definition a 'generalist' and therefore won't know everything there is to know about each speciality. Instead you need to rely on your clinical skills (and gut instinct) whilst recognizing your limitations and accept that you won't always know the answer.

Don't get me wrong, dealing with uncertainty in general practice doesn't always have to be difficult. It is made significantly easier by the fact that you can see the patient again either later that day or the following day, to see if their illness is progressing as you would expect. As a rather wise colleague once said to me 'time is your great ally'. Patients really don't mind coming back to see you for review, even if it is the same day, as it makes them feel that you are taking their problem seriously and that you care. The same applies to a phone call which most patients welcome and are often surprised at. In some circumstances it can be wise to share your uncertainty with the patient so that they understand that you are only human and therefore may not yet (if ever) be able to give them a definitive diagnosis for their symptoms. All of this is, of course, made easier in general practice by the fact that you may already know the patient's personality, background, and family, as well as their threshold for illness behaviour (and complaining!). These vital pieces of information enable you to put their symptoms into the context of their daily lives and will also help you to decide which ones are sick and which are just the 'worried well'.

Autonomy and decision-making—'standing on your own two feet'

Making your own decisions is something which takes a while to get used to in general practice when you have come from a hospital setting where you have no doubt been part of a large team. Whilst you will have been required to manage the initial care of patients that you have admitted in hospital, the decisions that you made were likely to have been reviewed by a more senior doctor within a few hours and eventually later that day by a consultant. This process makes for excellent learning as you begin to understand how and why decisions are made and whether or not you are on the right track when it comes to your diagnostic and management skills!

In general practice however, your decisions are your own and there is no one telling you what you should or shouldn't be doing (or if you're doing it right for that matter!). Instead you need to make a plan based on the clinical information available at the time and stick with it. Of course, you can always ask your trainer or another GP if you get stuck but essentially you are responsible for managing each of your patients yourself and can't just hand over the difficult ones to the next doctor on call for the day!

Whilst having this autonomy may seem scary at first, it really can be quite liberating to have the opportunity to trust your own clinical skills and act accordingly rather than do just what your registrar or consultant has told you to do.

Time management

No one is doubting the fact that hospital jobs are busy or suggesting that junior doctors sit around in the mess drinking coffee all day, but the way that you manage your time will be completely different in general practice. Having a list of patients to see in your surgery in some ways is similar to doing a ward round, except that you have to take the history, examine the patient, and make decisions all within a very short space of time. Think about the last time you clerked a patient whilst on call in the hospital. How long did it take you? And how many times did you go back and see that patient to review them and possibly change your management decisions? In general practice, you will have much less time in which to do all of this and will also probably have a queue of patients in the waiting room. Learning to consult in 10 minutes is a skill in itself and is something which takes months to achieve well.

Prescribing

As part of making decisions and having autonomy, it will be up to you to choose the class and dose of medication that you prescribe for your patients. You are likely to find yourself prescribing drugs that you may never have prescribed before or have only ever written on drug charts under the instruction of your consultant. Knowing what drug to start, when, and at what dose can take time to master which is why it is essential that you get to know your way around the *BNF*. Prescribing new medications for the first time can be daunting, but as long as you consult the *BNF* and ask your trainer if you're not sure then you shouldn't have any sleepless nights.

Patients will also often come in with queries regarding medications that were started in hospital and it's then that you will learn to appreciate the benefit of a well-written hospital discharge summary. Very often there will be no mention about why certain medications were stopped or started and you will find yourself trying to second guess what happened in hospital and what you as the GP will need to do next. Avoid simply re-authorizing medications that were started in hospital without checking doses and any monitoring that may be required for that drug.

Primary/secondary care interface

If you have only ever worked in a hospital, it can sometimes be hard to appreciate the complexity of the role that GPs play in their patients' lives and the negative stigma that is sometimes attached to primary care. Communication between primary and secondary care has historically always been slightly lacking, and this can often have negative consequences for the patients involved. Delayed or inadequate discharge summaries, stray clinic letters, and stroppy junior doctors on the end of the phone, can all make your life in general practice rather difficult. Of course, the hospital doctors aren't deliberately trying to cause you any hassle and in fact, in most circumstances it's the 'process' that fails. The communication between GPs and the hospital consultants really could be improved in most areas in order for the NHS to work better as a team. Hospital doctors don't understand what GPs do and GPs don't understand what hospital doctors do and both seem to be battling over whom is the most hard done by. The introduction of GPwSIs and clinical assistants in certain specialities may go some way towards solving this predicament but I suspect that it will be a long time before doctors in primary and secondary care are working together more cohesively.

This shouldn't, however, affect your decision to try and get hold of an appropriate consultant for advice when you need to. Remember that registrars and junior doctors are still training and their lack of confidence in dealing with your query may often come across as them being rude or unhelpful. Consultants are busy people too but it is not unacceptable to attempt to discuss difficult management decisions if it is in the best interests of your patient—after all that is what we signed up for when we qualified as doctors, looking after people.

Consultation models

No doubt you won't have been on the GP training scheme long before you come across the concept of consultation models[1]. There have been many different models written over the years in an attempt to improve doctors' consultation skills. They are essentially all very similar, but each with a slightly different focus. They aim to provide a structure to the consultation that is slightly different from that which you learnt in medical school and doesn't simply go along the lines of history, examination, and forming a management plan. Using a consultation model in your practice can not only help to improve your communication skills but can also make sure that both you and the patient feel more satisfied at the end of it. Often during your hospital days as a junior doctor it is more your agenda rather than the patient's that you strive to meet, as you try to keep up with the long medical ward rounds and the four-hour A&E breach targets. You will probably have at least one teaching session on consultation models as part of your VTS study days, but if you don't it's worth looking them up and trying out a few of these during your surgeries (in particular that written by Roger Neighbour[1] which follows a similar consultation structure to that of the Clinical Skills Assessment).

Safety netting

This concept created by Roger Neighbour in his consultation model, refers to the process of making sure that you have provided both the patient and yourself with a plan to manage the unexpected should it happen, in an attempt to deal with any uncertainty surrounding the diagnosis. Unlike in hospital medicine, in general practice the patient will not sit on a ward having regular observations by nursing staff in order to pick up any potential complications or deterioration in their condition. Instead, the patient needs to be aware of what things should prompt them to seek further medical attention and have a plan in place as to what they should do should this happen. This not only means that you sleep better at night (in the knowledge that even if things change or you haven't quite got the diagnosis right, then the patient should come to no harm) but also that the patient feels reassured by having been given appropriate instructions as to what to do should their illness not progress as expected.

Investigations

Requesting investigations in primary care should be done with careful thought and consideration of what you expect the results to show. Doing 'blanket tests' on patients not only creates extra work but can also lead to the finding of inconsequential abnormalities which are difficult to ignore once they have been detected (see **CHAPTER 11, 'YOUR FIRST GPST POST AND SOME BASIC TIPS ON HOW TO SURVIVE'**). Unlike in hospital, the results can take anything from days to weeks to be available and therefore you will need to decide on your management plan without this information in the first instance. You cannot simply repeat them on a daily basis to watch trends and monitor abnormalities like you can in hospital when all you need to do is fill out a blood form and leave it for the phlebotomists to collect. Remember also that having to go and get a blood test or x-ray done is likely to be very inconvenient for the patient as it will probably require a trip to the local hospital or another clinic and could prove almost impossible for those with transport difficulties or disabilities. Think carefully about whom you send for blood tests and why, and remember that it will be you and not your registrar or consultant making decisions regarding the results.

Working alone

Working alone has many benefits, including the opportunity for working autonomously and at your own pace; however, if you're used to the hustle and bustle of a busy hospital environment then you might find general practice quite lonely and isolating. Of course this depends largely on the size and type of practice in which you work and what type of personality you are. If you work in a small practice there may be some days that you do not see a single person other than your trainer (if that) and even if you work in a large practice, if the practice doesn't have a communal area where members of staff meet at meal or break times you could be in the same situation. This differs enormously from working in a hospital environment where you tend to spend most of the day working in teams and even if you are on call and working alone, you will be surrounded by nursing and other hospital staff. If you're a sociable person, this change in environment can really be quite difficult to get used to and can sometimes leave you feeling isolated and lonely if you aren't prepared for it. My fondest memories of working as a junior hospital doctor are the lunchtime 'catch ups' in the doctors' mess and the 'tea and toast' parties at 5am whilst working in A&E. If you feel that you might find yourself getting lonely during your GP posts make an effort to spend time with the nursing and admin staff at the practice during lunchtimes and make the most of your VTS study sessions where you get to meet up with other trainees.

Continuity of care

This is probably the biggest difference of all when comparing primary and secondary care. Whilst you may develop a relationship with some of the more 'long-stay' patients that you look after on the wards, the chances are that most patients you see will have either been discharged or moved to the next ward within a few days. You will probably never see them again and therefore never know the outcome of their illness. This can be particularly true when you are on call and go home before the patient has been reviewed by the consultant. If this is the case then you don't even benefit from finding out whether your assessment, diagnosis, and management of the patient were correct, which can be frustrating in terms of your learning.

In general practice, however, you will undoubtedly see some patients on a fairly regular basis and the quality of the doctor–patient relationship that you have with them will affect the outcome of the consultation. Trying to maintain a positive doctor–patient relationship often affects the decisions you make when managing certain difficult situations, as a breakdown in this relationship can not only make subsequent consultations very difficult, but if that patient decides to leave the practice, they may also take with them their friends and family who may be registered on your list.

However, establishing a good doctor–patient relationship with your patients is both rewarding and satisfying. Being able to provide that continuity of care means a great deal to our patients, in particular to those with chronic or terminal illnesses, or in fact children who you are able to watch grow from being toddlers to teenagers (if you're at the same practice long enough!). Within each patient's life, they may meet a whole host of different doctors and medical staff, but it's the GP that tends to remain constant and provide support both medically and socially throughout this time and I believe that to be both an honour and a privilege.

> 'The BEST part of GP training is the GP consultation—a fascinating exchange between two people with different agendas. The nuances of the consultation are great fun.'
>
> ST2

It's not until you have spent time working in general practice that you will really begin to appreciate how different hospital medicine is from primary care, but it won't be long before you too are moaning about inadequate discharge summaries and elusive consultants!

REFERENCE

1 Simon C, Everitt H, and van Dorp F (2010). Consultation models. *Oxford Handbook of General Practice* 3rd edn, pp. 90–1. Oxford University Press, Oxford.

CHAPTER 11

Your first GPST post and some basic tips on how to survive!

Introduction

For many trainees, the thought of starting their first GP post fills them with utter dread and this is normally attributed to the common theory of 'fear of the unknown'. Whilst some of you may have been lucky enough to have completed a four- or six-month placement in general practice during your foundation years, there may still be some people out there whose only experience of general practice was as a medical student, which was not only likely to have been a long time ago but may well have been back in the days when general practice was very different.

This chapter therefore aims to answer many of the questions that may be keeping you awake at night in the run up to your first 'practice posting' and should also provide you with most of the information that you need to ensure that your first few weeks in the post run smoothly.

Before you start

What will I need to have done in advance?

The pre-GP post checklist:

- Find out where your surgery is—do a 'dummy run' of getting there from home. Turning up late on your first day doesn't go down well with your trainer
- Make sure that you are on the Performers List—see the following question **'WHAT IS THE PERFORMERS LIST?'**
- Have a chat to the ST who has just left—they should hopefully give you any essential inside info that you should know before you start
- Take any necessary or requested documents in to the practice manager, e.g. driving licence, passport, and certificates (although some may just ask you to bring these with you on the first day)
- Update your medical defence organization status to that of a GP Registrar (or equivalent). This costs considerably more than being a hospital doctor but the extra component can be claimed back from the PCT via your practice manager. You will have to pay it in advance though (although most offer monthly direct debit) and will get the reimbursement back at a later stage depending on how your practice manager organizes this. For more information on this see **CHAPTER 7, 'THE MONEY STUFF'**

> Tip: Make friends with the practice manager early on. If they like you they can make your life so much easier when it comes to annual leave, filling in forms, sorting out appointments, and filling in the right QOF codes! Oh yes... and most importantly they'll also make sure you get paid on time!

> Tip: Buy yourself a notepad or decent-sized diary at the start of your GP placement to make a note of any complex patients or gaps in your knowledge that you may come across during your surgery. You can also use it to make a list of referrals that you have sent so that you remember to chase up the outcome at a later date.

What is the Performers List?

The Performers List is a list of GPs within your PCT that are eligible to work within primary care. It was created in 2004 as a way of protecting NHS patients and services by enabling the NHS to better regulate practitioners. In order to work as an ST in general practice you must be a member of your local PCT Performers List before you can see any patients. This is really important as it is also a prerequisite for your medical defence organization, who will not cover you to work as a GP if you are not registered on a list. Getting yourself on the Performers List can normally be organized with the help of the practice manager at the surgery where you are hoping to work. They should be able to give you all of the necessary forms to fill in and advise you on what to do next, but if you are stuck then make sure you contact the PCT and ask for some assistance. Essentially the responsibility lies with you to get your name on the list and you must therefore make reasonable attempts to do so.

If you have done any general practice in your foundation years and were on a Performers List at this time, contact your local PCT to check that still stands, as some PCTs will remove you from their list on the specific date that your GP placement ends. Never assume that you are still on the list and it is your responsibility to check this in advance.

You are only allowed to be on one Performers List at a time, so if you were previously registered with an alternative PCT then you will need to withdraw from this one and reapply to the new one. The good news is that you can still work whilst this is going through if you are simply moving from one PCT to another unlike when you are applying to be on a Performers List for the first time.

> Tip: Think about applying to be on the Performers List around three months before you are due to start your GP placement as it can be a lengthy process and you cannot work in general practice unless you are on it.

Each PCT is different, but in general, in order to get on the Performers List you will need to:

1. Complete an application form detailing your training/experience to date
2. Make sure that you have an up-to-date CRB inclusive of vulnerable adults (and if not complete an application form for one)
3. Supply them with names of two references who will be contacted by the PCT on your behalf

Depending on the efficiency of your local PCT Registration Service this process can take anything from three weeks to three months, so make sure that you get this sorted well in advance. If you don't manage to get on the Performers List in time for your placement then you will not be allowed to work and will subsequently not get paid.

There is talk on the horizon that there may be a single Performers List to which you apply at the start of your training that will cover you for the duration (so that you don't have to worry about re-applying and checking your status in advance of further

GPST posts) but there is nothing set in stone at present. I would suggest, however, that you keep your eyes peeled for any changes in this process so that you aren't caught out by any of the new rules and regulations.

Will I need to buy my own equipment in advance?

No, as long as you have your own stethoscope you should be fine. Your practice should provide you with all of the equipment that you need and if they don't then perhaps discuss this with either your trainer or practice manager. Most practices will have a designated registrar bag that you can use although you will probably be expected to have your own stethoscope.

In terms of buying equipment it's perhaps best to wait until you're in the job before you decide what items you need and what you would actually like to buy. You will eventually need to accumulate your own doctor's bag equipment but you have plenty of time to do this. The training years are expensive enough without trying to find the money to buy a sats monitor or ophthalmoscope.

How many hours will I work?

Most surgeries open between 8.00 and 9.00am and close between 4.30 and 6.30pm. If your surgery operates an extended hours policy, it may open as early as 7.00am and close as late as 8.30pm on some days (and may also be open on Saturday mornings!). You are contracted to work 10 sessions, each of four hours duration and usually split in the following way[2]:

- Seven clinical sessions (28 hours)
- Two structured educational sessions, e.g. VTS/half day release (eight hours)
- One independent educational session, e.g. tutorials/debriefing after surgery (four hours)

However, this split should be flexible and ideally based on your educational needs. Remember that in order to succeed in general practice you need to see lots of patients and worrying about how many clinical versus educational hours you are doing can often be distracting from the end target which is qualifying as a good GP. If you are asked to work during an extended hours surgery this should be instead of, not as well as, the 40 hours that you are already doing. You should also make sure that you are properly supervised during these additional hours and should not find yourself working late in the surgery alone.

Do I need to have a car?

Officially no, but in real life yes. According to the contract 'you must either hold a current valid driving licence' or have 'use of a motor vehicle, or provide alternative means of offering emergency and domiciliary care to fulfil the requirements of the post'[2].

You will also need to notify your insurance company that you will be using your car for business purposes otherwise you might not be insured when you crash your car into old Mrs Smith's unfortunately placed rockery. Most insurance companies do not charge any extra for this so you've nothing to gain by not doing so. Make sure that if possible you have a car that is vaguely reliable too. Breaking down on your way to work on a regular basis can be quite frustrating both for you, the practice, and your patients; in the same way that being stranded in the middle of nowhere whilst trying to do a home visit would be rather inconvenient.

Do I need a mobile phone?

Yes but very few surgeries will provide you with one so this is something that you are responsible for organizing. Whilst you shouldn't be required to make too many outgoing calls, it's particularly useful to have a mobile phone when you're out on visits in case you need to phone your trainer for advice or more importantly ring for an ambulance. If you are included in your surgery's on-call rota then you will also need to be contactable in an emergency outside the normal opening hours. Being contactable by telephone also forms part of the educational contract.

On the day

Where do I go on my first day?

This will obviously depend on where you are working and whether or not your surgery has issued you with an induction timetable. Most trainees will meet with the practice manager at the surgery on the first day, who will show them around the building and introduce them to the other members of staff. You may not get to meet your trainer or the other partners in the practice until later that morning once surgery has finished. Hopefully you will have already met most of the staff either during your informal look around or when you dropped in to sort out bits and bobs before starting the post. If you haven't already done so you will need to provide the practice manager with evidence of your GMC certificate, medical defence certificate, driving licence, passport, and primary medical qualification certificate at the very least. Some will also ask for evidence of your hepatitus B status. Keeping all of these items together in a folder makes you less likely to forget anything and more importantly less likely to lose anything!

What will I do on my first day?

Probably not a lot. You are likely to spend the first day sorting out paperwork, meeting the team, and perhaps sitting in with one of the regular qualified GPs. This all depends on what sort of induction timetable your trainer has planned for you (if any—some are better at this than others!).

Should I have an induction timetable?

Yes, ideally you should. This saves you hanging around like a spare part feeling like you're getting in the way for the first two weeks and also means that the practice can make sure that you are familiar with all of the policies and safety procedures before you start. It's always useful within the first week to spend a few hours within each area of the surgery to get an idea of what they do and how things work, e.g. nurses, dispensary, reception, health visitors, district nurses, and midwives. It may be worth asking the practice manager or your trainer about your induction timetable, when you meet or contact them prior to the post.

Will I get to sit in with the regular GPs?

Hopefully yes. The induction period is an excellent opportunity to learn from your seniors and watch the ways in which they consult. Every GP is different and watching partners who work very differently will emphasize how there isn't a right or wrong way to skin a cat. Often different consultation methods will still yield the same results and this can be fascinating to observe. Sitting in with partners or salaried GPs also helps you to soak up the general ethos of the practice and the way in which things work.

Try not to switch off when you are sitting in during these times. I remember very clearly back at medical school sitting in with consultants in clinics and regularly having to kick my clinical partner under his chair to try and prevent him falling asleep during some of the more boring neurology clinics. One day I completely failed in my task however and he only came around after he woke himself up snoring—all very embarrassing for everyone involved.

Whilst you may be fortunate enough to get to sit in with qualified GPs at other stages of your training, the induction period really is the best time to make the most of doing this, to see what consultation skills you do or don't want to consider putting into practice yourself.

What computer system will I use? Will it be the same in every practice?

Each GP practice will use its own computer package, but the five most popular ones used are:

- EMIS LV
- EMIS PCS
- VISION
- PREMIER SYNERGY
- SYSTEM ONE

Each one operates quite differently and appears very differently on the screen. EMIS LV uses mostly computer keys rather than the mouse and requires you to pick what you would like to do from coded options lists. EMIS PCS is similar but has more of a Windows feel to it and relies on the mouse buttons more. VISION works in a similar way to Windows allowing you to use the right and left click of the mouse to move across the screen and input data. PREMIER SYNERGY is very similar to VISION but displays the information in a slightly different way. This also relies more on the mouse than on the computer keys but there are also shortcut keys that can be used to save time. Essentially whichever one you get used to first of all will probably stay as your favourite, although being able to use other packages comes in very handy once you qualify and are looking for locum work.

All packages require you to code essential information for QOF purposes, although the QOF data may be input in different ways and via alternative packages. Whilst this is important to be aware of, in your first few weeks you should concentrate on being able to accurately input your clinical data rather than worrying too much about QOF codes.

Who will show me how to use the computer?

This is most likely to be either the practice manager or the lead IT person in the administration department of a bigger practice. Each practice should have a booklet on how to use the package and important Read codes that you should be aware of. Whilst you can be shown the basics of how to use each package, there really is nothing better than just spending an afternoon playing around with the notes of a mock patient and trying to input different types of information. At least this way you can make as many mistakes as you like and it really doesn't matter. It's also worth chatting to your trainer and other GPs within the practice to see if they have a specific way in which they like the clinical information to be added to the patient's notes. Some surgeries prefer the information to be added by 'symptom' headings, whereas others are keener for the problems to be listed under 'diagnostic' headings.

What are Read codes?

Read codes are specific codes used by each of the computer systems to document symptoms, signs, diagnoses, and management options in a patient's computer records in order that they can be used to audit clinical information and search for individual clinical details within a patient's notes. They are named after a GP called James Read who created and sold the concept to GP land for millions of pounds. They are also used to document essential information required for QOF which is why you will hear partners harping on about them so much (correct Read codes = correct QOF = more money).

What is QOF?

QOF stands for Quality and Outcomes Framework and essentially forms part of the way in which GP practices get paid for what they do.

Will I have my own room?

Hopefully yes. Although if your surgery has part-time staff or more than one training doctor at any one time, then you may have to 'hot desk' (work wherever there is a space!). 'Hot desking' isn't ideal as it's difficult to get settled and feel 'at home' in someone else's room, but at times it is unavoidable. If you do have to share a room, try and discuss this early on with the person that you're sharing with so that you can come to a mutually acceptable agreement in terms of the way the room is left and what equipment will be available.

> Tip: In the first few weeks make the most of having some time to get your room sorted. Make a list of useful phone numbers and protocols, organize your drawers with the appropriate forms, and make sure you have all of the clinical equipment that you require.

Seeing patients

When will I start seeing patients?

This varies considerably depending on the size of the practice, your trainer, the facilities available, and your ability. If this isn't the first time that you have worked within a general practice setting then you may start seeing patients fairly early on. If, however, you have only ever worked in a hospital environment, then you may need easing in a little more gently as it really is very different. Even little things like getting used to operating a tannoy can be scary at first!

Essentially you can start seeing patients when you feel you are ready, but it's likely that you will spend some time sitting in with other GPs first, not only to get a feel for the place and for working as a GP in general, but so that you also become familiar with the computer system before being left alone trying to fathom it when you have patients in the room. Perhaps discuss this with your trainer on the first day so that you know what is expected of you and when. Some trainers like to push their trainees early and 'drop them in at the deep end' as it were, whereas others like to ease you in gently and essentially it's about coming to a joint decision on what is best for you. Dropping someone in at the deep end can cause terrible knocks to the confidence when they struggle to keep up, whereas easing someone in too gently can mean that they become arrogant, lazy, or even bored.

How long will my appointments be?

When you first start seeing patients, each appointment should be 20–30 minutes long, in order to give you enough time to get used to using the computer and familiarize yourself with the different forms and lab bottles that you may need to complete. You will also find that actually taking a history and examining the patient may take longer than you thought it would, as you may still be very much in 'hospital mode' and it takes some time before you break out of that mould.

The plan should be that as the weeks to months progress, your appointment times should start to shorten, with perhaps less and less catch-up gaps included (although don't try and run before you can walk, as running very behind is neither good for the patients nor you and your self-esteem). Make sure that you are having regular discussions with your trainer about the pace at which you would like to shorten your appointments or remove any catch-up breaks. If you do shorten your appointment times but then start to struggle, there is no harm in adding a few extra breaks back in to make sure that you are surviving. You don't win any prizes for seeing 30 patient in three hours if you then require a bottle of vodka to calm you down once you get home, in the same way that you won't make a good impression if you are still wanting to have 20-minute appointments six months in, when it is clear that you are consulting more quickly than this and in fact are just being lazy.

> Tip: Make sure that you are having regular discussions with your trainer about the pace at which you would like to shorten your appointments or remove any catch-up breaks.

By the end of your first placement you should aim to be consulting comfortably at 15-minute intervals without running too far behind. However, everyone is different and some may find that they are easily managing 10-minute appointments by this stage, although try to make the most of the training time available rather than rushing your consultations just to boast to your peers that you are consulting in 10 minutes.

How many patients will I see per day?

This will depend on how long your appointments are, what time they start and finish, and whether or not you have any breaks in between. Some surgeries have a break mid-morning and mid-afternoon for coffee and others work straight through. When you first start you will probably only see four or five patients in each morning or afternoon session and whilst this seems like hardly any, it's amazing how long it can take you to get to grips with the computer and work out how to fill in all of the different forms, not to mention finding out where they are kept. As you progress and your appointment times get shorter you will probably ending up seeing between eight and 16 patients per session with perhaps more on days that you are duty doctor or on call. But don't panic. This won't all happen overnight and there should be a smooth transition to seeing these sorts of numbers. Make sure that you are communicating with your trainer at each step of the way; otherwise you may feel that you are doing more than you can cope with and this will only knock your confidence and lead to resentment.

Who can I ask for help? (E.g. in the middle of surgery)

Ideally your first port of call should be your trainer if they are around. Some GP trainers will have longer appointments during times that they are supervising you in order to accommodate any help you require; however, others will have surgeries as normal and will fit in your queries around their patient list. Perhaps have a chat with your trainer before you start about the way in which they would like you to approach them for help. Some may be happy for you to call them in the middle of their surgery, whereas others may prefer that you wait outside their door until the patient that they are seeing has left. Some also have an on-screen message system that enables the trainee to highlight to their trainer that they need advice. If the problem is non-urgent and doesn't need an immediate answer, you may be asked to just keep a list during your surgery that you can then discuss with your trainer at the end of that session. Essentially, every trainer is different and will have a preferred method of helping out. The most important thing to remember is that they are there to help you and you shouldn't feel guilty for asking when you get stuck. If you ask most GP trainers they will say that they worry more about those trainees that don't ask than those who do. They would much prefer that you are able to recognize your limitations and ask for help than struggle on in silence making possibly risky or dangerous decisions about patient care. Make sure, however, that when you approach them you have done the basics and are clear about what sort of help you require, e.g. do you need them to come and see the patient with you or are you happy to get some advice over the phone? Some trainers will want to come and see the patient with you regardless of this but at least if you let them know in advance then they can decide on their preferred approach. Learning which patients you do and don't need to discuss in the middle of the surgery often takes time and you may find that in the early stages you are asking frequently about minor issues. Please don't worry about this as it is all part of the learning process and things will get easier as time goes on.

> 'My trainer was excellent and never made me feel like a pain for interrupting her surgery when I had a problem. In the early days, she was also more than happy to go through all of the patients that I'd seen in that session to make sure that I felt happy with what I had done. This was really important in helping me to build my confidence and meant that I didn't go home and worry about patients that I'd seen for fear of having done the wrong thing.'
>
> ST1

What if I need help urgently?

This may sound like a silly question, but you may have never experienced a situation before where you are alone behind a closed door with a patient down a long corridor. Hospital settings normally feel pretty safe as there tends to be plenty of people around and instant help is usually only a shout away. However, in general practice it's important that you know how to summon urgent help should the patient become aggressive or acutely unwell, as you may not have time or be able to run out of the room to escape or get assistance. Most surgeries have a panic button either on the telephone or computer screen and you may not get told about it unless you ask. Very often these buttons are pressed by accident causing swarms of practice staff to come running to your room, but don't worry about it if you set if off by accident, as in these circumstances it's reassuring to know that the system works! On the same subject, think carefully about how you arrange your room. Try to avoid sitting with the patient between you and the door as should the patient become violent, you need to be able to escape without any obstructions. This all sounds very scary and the chances of you ever having to need to worry about it are slim but it's better to be safe than sorry!

How do I know which drugs I should prescribe?

By this question what I am referring to is not what drugs should be used to treat what but instead what drugs **your practice** use to treat what, in terms of your practice formulary. Even if you do not have an on-site dispensary at your surgery there will still be certain drugs or brands of drugs that your practice will prefer you to prescribe based on your local dispensing guidelines. Make sure that you have a copy of any formulary that the practice may have, so that you are aware of some of the cost differences between different drug brands and any QOF prescribing incentives that your practice are currently adhering to. Spending a morning with one of the dispensing team, if you have one, is invaluable and gives an interesting insight into drug costs, budgets, and dispensing discounts and deals.

When will I start having to do home visits?

The prospect of doing home visits is initially a very scary one to most trainees starting out in their GP career. Having worked in the comfort of a hospital setting and most probably within a team for the last two years, being sat in a room on your own is bad enough, let alone being sat in someone else's home on your own and being expected to know what you are doing. Home visits are actually never as bad as you think they are going to be. There is a real knack to having a successful home visit and it's something that takes years to perfect. Getting over the whole concept of leaving the surgery and getting out on your own in the car is just the first step in the battle, but in time you will learn that home visits in the main can be quite satisfying and productive.

Each surgery will have their own policy on home visiting and some are stricter than others. Rural practices with an elderly population may have historically done lots of home visiting leading to raised patient expectations and making it very difficult to change. The elderly in these areas may know no other experience than to have the local doctor visit them at home, despite the fact that they manage to get down to Morrisons on a Friday and drive to bingo on a Tuesday night. Other surgeries will have a very strict policy on visiting and will only visit those who are either house-bound or terminally ill. Very few practices visit children these days as there is little to be gained from doing this, although some parents will still request it and be shocked when they are asked to bring the child down to the surgery to be seen instead.

It is unlikely that you will be expected to visit patients alone within your first few weeks of starting the post, as you will only just be getting to grips with some of the more basic GP principles, without wanting to worry about navigating your way around some country lanes with only a stethoscope and tongue depressor to call your doctor's bag. Often practices will send you out with the qualified GPs on their visits in the first few weeks to get a feel for what it's like and then after that you might start off by going to see some of the more straightforward cases that shouldn't cause you too much distress.

The first time that you go out alone can be frightening with the most common fears being that the patient is either going to be really sick or that you will see them and assess them and still have no idea what is going on and what to do. In these cases always remember to go back to first principles:

- Is the patient sick and if so are they sick enough to require hospital admission? If yes, discuss this with the patient and call an ambulance if appropriate.
- Is the patient unwell but not acutely sick? In these circumstances you have more time than you think to decide what to do. Your trainer (or another accessible GP at the practice) should always be on the other end of the phone to discuss difficult cases with you and no one will think badly of you for recognizing your limitations and doing this.

In the majority of cases the home visits are neither scary nor complex and more commonly involve going out to see elderly or housebound patients with minor illnesses or difficult social problems. Managing the social problems can, of course, be more of a headache than managing the acutely ill patients, but there tends to be less urgency in sorting these issues and instead more leg work and phoning around. Patients in general are very grateful for you taking the time to visit them at home and will often offer you cups of tea and cake, completely oblivious to the mountain of paperwork that you will have to do on your return to the surgery whilst they catch up on this week's 'Diagnosis Murder' after you've left.

How is the home visiting divided up?

This varies between practices and in general is either done during the morning coffee break or is organized by the duty doctor for that day. Very often, individual GPs will pick and chose patients that they know well or see regularly, so as the ST you will either be given the more straightforward ones that don't need to see anyone in particular or alternatively the really awful difficult ones that the regular GPs can no longer face going to see. If this is the case it may be packaged up with a sparkly gold bow by your trainer, suggesting that it would be a 'good experience' for you, when what they actually mean is that from experience they would rather eat their own toenails than have to sit on that patients urine-stained couch and give them another enema for their recurrent constipation. However, think positively about these situations. One day you too will be the fully fledged GP who has trainees to do some of the less appealing jobs, and let's face it, don't pretend you didn't ever farm out some of the nasty jobs to the FY1 when you had progressed to being a FY2!

The paperwork stuff

Will I have to sign repeat prescriptions?

Initially no. Repeat prescribing can be difficult to get your head around and unless you are vaguely familiar with the patients and the medications that they are on then it can be a little daunting. As time goes on however, these sorts of things should be gradually phased in, as it all forms part of the job and whilst tedious and time-consuming, it's important that you know what you are doing.

In surgeries where they have a dispensary, the prescriptions may go straight through to the dispensary from your computer, in which case you will need to make sure that you sign these prescriptions at the end of each session. Some surgeries may also expect you to sign prescriptions issued by other doctors so that they are not left unsigned for days on end. Be careful doing this in the early days as by signing these prescriptions you are assuming the responsibility for prescribing the drug and this could come back and bite you if you don't take this seriously.

Will I have to analyse blood results?

Ideally yes, although at first your results may well be filtered by your trainer to make sure that they are filed correctly and dealt with appropriately. It is good medical practice for results to be reviewed and acted upon by the doctor that requested them, as it can be very difficult to interpret an abnormal blood result without having a grip on the background clinical picture. Each computer system has their own way of downloading and assigning blood results to individual doctors and its worth familiarizing yourself with this early on so that you can get used to dealing with your own blood results right from the start. Knowing what to do with abnormal results in primary care is quite different from that in the hospital setting. Unlike on the ward, you can't just check them again tomorrow if they're slightly abnormal and essentially the decision you make is yours and not that of your registrar or consultant. You also may have waited anything from a few days to a week to get the information back and so the clinical picture of the patient may have changed since the time the bloods were taken. As is the case in hospitals though, make sure that you are treating the patient and not the blood results. An otherwise healthy 25-year-old whose routine bloods reveal a potassium of 6.5, is probably just unfortunate enough to have had their sample delayed in the lab leading to a haemolysed specimen, rather than be acutely unwell with hyperkalaemia of unknown origin. In these circumstances, make sure that you call the patient first and see how they are before you anxiously ring for an ambulance to take them straight from work to hospital in a panic.

Learning how to manage abnormal blood results in the community is not easy and takes time and relies on experience. The grossly abnormal results are often easy to sort, but it's the ones that are just mildly awry that cause the most anguish and make you wish you had never requested them! Moral of the story? Only request those blood tests that will discriminate between your differential diagnoses—although this is easier said than done and even experienced GPs are guilty of requesting unnecessary bloods from time to time!

How do I request a blood test/x-ray?

Find this out early on in your placement. Scrabbling around trying to find the right forms is not only embarrassing in front of the patient but also takes up valuable time. Most forms can be completed after the patient has left but depending on your local policy you are likely to need to complete blood and x-ray forms with the patient so that they can take them away. Some surgeries have label machines which prevent you having to hand-write all of the patient details and others will have a form that you can access within the patients notes which automatically inputs their details. Both of these save considerable time during the consultation so make sure you enquire about them early on!

> 'It wasn't until the last month of my GP placement that I realized that the grey box on my desk with the green light was in fact a label machine. I could have saved hours by not having to hand-write all the blood forms!'
>
> ST2

Will I have to deal with any clinical letters?

Yes, but ideally they should only be those letters which are relevant to you and the patients that you have seen. Some hospitals will write back to the referring doctor whereas others will simply address their letters to the default named GP on their computer system. This can mean that post which should have been addressed to you will often bypass you and be seen by another doctor within the practice. It's therefore well worth keeping a note of any patients that you have referred and when, and checking their notes every so often to see if they have been seen by the specialist yet and what the outcome was. Unless you make a conscious effort to look out for letters written about patients that you have referred, then you may well never see them and will therefore miss out on the invaluable experience that is reflection.

Will I have to complete insurance reports and medicals?

Completing insurance forms and medicals is not really part of your job description and in fact according to the suggested BMA contract you should 'not be required to perform duties which will result in the receipt by the practice of private income, unless an arrangement to the contrary is entered into before the commencement of your GP Specialty Registrarship'[2]. Most of these forms are carried out in addition to the NHS services which the surgery are expected to provide and therefore incur a significant charge to the patient (which is then kept by the practice regardless of who competed it!). The same applies to cremation forms, the money received from which is normally kept by the practice.

However, completing these types of paperwork will form part of your everyday duties once you qualify and if you have never had the experience of doing them as a trainee, then you may struggle once out in the big wide world on your own. Make the most of any opportunities which arise to complete these types of paperwork on patients that you know, as this is much easier than doing the same thing for a complete stranger (and sometimes you may have been the only doctor in the practice to have ever seen the patient). Perhaps then use the report as the basis of a case-based discussion or tutorial with your trainer so that it becomes an educational task rather than one of service provision. If, however, you are being asked to complete these sorts of things on a regular basis with no real obvious educational benefit then discuss this with your trainer. They are time-consuming and can be quite difficult, so if required to carry them out make sure that you are also allowed sufficient time to do so. It is, however, worth mentioning that whilst you will probably not be expected to do this in your ST2 year, it may well be expected of you in ST3 for your own educational benefit.

Study leave and teaching

How many days study leave am I allowed?

This varies depending on which deanery you are working for and is referred to by most as an allowance rather than an entitlement. Most deaneries permit 30 days of study leave allowance per year, divided up pro rata for six-month posts. These 30 days normally include the compulsory half-day or full-day VTS release sessions organized locally but not the one session per week allocated to planned educational study leave. The minimal attendance at the compulsory teaching is 80% which allows for annual and sick leave and if there is no teaching organized for a particular week then trainees are expected to either do a clinical session or an alternative educational activity.

Exam leave can be requested in addition to the 30 days and those trainees who are on a local or national committee are also entitled to an extra five days of professional study leave, in order to attend essential meetings and seminars.

Most deaneries will only allow five days per year of private study within the trainee's 30-day allowance and it must be agreed in advance with the ES to confirm that it is appropriate and relevant. Other deaneries, however, do not support applications for ANY private study leave and even discourage trainees from attempting external exams such as the DRCOG.

How do I organize my study leave?

First of all you need to check the criteria within your deanery for study leave eligibility. They are likely to have specific guidelines on what courses or study days you may attend and most are unlikely to fund any activities abroad. Once you have checked this you will need to fill out the appropriate form, available either from your deanery or the trust within which you are working if in a hospital post. Either way you will be expected to do this about six to eight weeks before the date that the study leave is required otherwise you may not be granted permission. If in a hospital post you may also have to arrange appropriate cover with your colleagues for your daily duties, so that the department is not left short. Think about this well in advance so that it doesn't prevent you from having the leave. You will then need to get the form signed by your CS to confirm that they approve the leave. For those in GP placements, you shouldn't have to worry about arranging cover although if your surgery has more than one ST then you are unlikely to be allowed to be off at the same time in the same way as for annual leave, although this varies considerably from practice to practice.

Can I get any study leave money—if so from where and how do I apply?

Yes you can but there is a limit which varies depending on which VTS scheme or hospital trust that you are working for. For those trainees in hospital posts their study leave budget is accessed via the hospital that they are working for, whereas those in general practice will need to organize reimbursement via their VTS course organizers. Accessing study leave money whilst working in general practice depends on how much is available and whether or not it is accessible at the time you apply. Some VTS schemes may distribute the funds at the end of the year once all of the applications have been processed. In general, ST3s should get around £400 per year although this may be less if your VTS scheme deducts certain amounts for locally organized teaching or courses.

How often will I have teaching with either my educational supervisor or someone else?

This will depend on at what stage of your training you are and what your educational needs are. Ideally you should have a total of one four-hour session per week of in-house teaching whether that is with your ES or another member of the practice team. This can either take the form of a tutorial, completing one of the workplace-based assessment tools or may involve spending time with other members of the practice team learning new skills. Your educational time should ideally be detailed on your regular weekly timetable although there should be some flexibility in this. Often just spending an hour after your surgery going through the patients that you have seen can be just as educationally beneficial as having a formal sit down tutorial on a given subject, so try to incorporate a mixture of activities. Essentially the time is yours so make the most of it and try to think in advance of what you would like to do during your designated sessions. Remember, you are no longer a spoon-fed medical student and instead need to focus your teaching and learning on gaps in your knowledge that need to be filled.

Common concerns

Will I have to work extended hours?

'Extended hours' differs to 'out of hours' in that it is a service provided by your own practice to your own patients in order to qualify for an enhanced service payment. Out of hours is very different and is normally run from a central location by a different group of doctors in order to cover a number of surgeries past a certain time. As part of your GP training you are required to complete at least 72 hours of true 'out of hours' work, whereas whether you will do extended hours or not depends entirely on how your surgery runs. For more information on this see the previous question **'HOW MANY HOURS WILL I WORK?'**.

Will I ever be left on my own in the surgery without a GP to ask for help?

Yes you might be, but that should not happen on a regular basis and should only really be in exceptional circumstances. Occasionally GPs will need to pop out of the surgery for visits or other commitments, leaving you alone in the practice but this is less than ideal and should only be for short periods of time. They should also make sure that they are contactable by phone so that you can at least phone them for advice. Being left alone tends to happen more frequently in smaller surgeries, where there are fewer people around to start with and staff sickness can often lead to circumstances when being left alone is unavoidable. This doesn't, however, make it right and if it is happening on a regular basis then you should discuss this with your trainer or VTS organizers. The BMA GPST contract[2] states that whilst 'you are an integral part of the practice team' you should still be 'supernumerary to the workforce of the practice' and 'at no point should the effective running of the practice be dependent on your attendance'. You should also 'not be used as a substitute for a locum in the practice'. The contract also comments that 'the trainer/educational supervisor, a nominated GP partner or other appropriate GP principal or salaried GP, undertakes to be available for advice, either personally or through a nominated partner, when you are on duty'[2].

What do I do if my trainer is away?

If you're used to only going to your trainer for help or advice then the prospect of them going on annual leave or study leave may be daunting. Some trainees find that they are already asking other members of the team for help, but if you haven't done this before then make sure that you chat to the other GPs within the surgery beforehand to find out who is the best person to ask for help whilst your trainer isn't around.

If all else fails there's always 'GPnotebook'[3]—a familiar friend to GP trainees across the country, which has got many a struggling trainee out of a sticky situation during a consultation.

When and how often will I have to be on call?

This entirely depends on the type of on-call system that your surgery runs. Some have an on-call or duty doctor for the day whereas others split the day in two so that you are only responsible for either the morning or the afternoon. Whether you will have pre-booked appointments during the time that you are officially 'on call' will again vary from place to place. Being on call normally means that you are responsible for dealing with any emergencies or additional queries and perhaps organizing other admin tasks such as assigning visits and dealing with urgent paperwork. It can be quite stressful in the early stages and should therefore not be something that you are asked to do in the first few weeks (unless very closely supported by your trainer) and instead should be commenced when you and your trainer are in agreement that you are ready for this. It is good practice for there to always be a senior GP covering you whilst on call and for them to be easily contactable either within the surgery or by telephone. How often you're on call will depend on the size of your surgery and how many doctors you have available, but you shouldn't be doing more than any of the qualified GPs. Different surgeries will expect different things of their trainees, with some giving more responsibility and on-call commitment than others. If you're not sure what is and isn't fair to be expected of you then discuss this with your peers or your VTS course organizers.

Do I need to ask my trainer's permission before I can refer a patient to secondary care?

In the early stages your trainer may well prefer that you discuss all referrals with them before sending them as there may be another way in which the problem could be managed within the primary care setting. Research has shown that trainees will often err on the side of caution when it comes to referring onwards at the beginning of their GP placements and it may well be that a more senior experienced GP can offer help and assistance instead. This is true particularly within larger practices where there are often GPs with special interests in different fields who may offer an opinion as good as or equal to that of someone in secondary care.

Discuss making referrals with your trainer in your initial few meetings to try and ascertain what their preference would be. As you grow in confidence and spend more time in general practice then the need to 'run it past' your trainer slowly ebbs away, but it can take time and it needs to happen at a pace that is acceptable to both you as the trainee and your trainer. Auditing the referrals that you send can also form the basis of a tutorial and help to identify some of your learning needs.

Tip: As the GPST you will probably get given copies of the new NICE guidelines by the practice as they are produced. Instead of using them to mop up your coffee spills, read through them as they arrive and make notes of salient points—these will serve you well as revision notes when you come to do AKT.

Summary

Your first post working in general practice can be a very steep learning curve, but it is exciting and rewarding and should be thoroughly enjoyed. If this isn't the case, then don't be afraid to speak up and discuss how you feel with your trainer. It's much easier to work these things out in the early stages than having to try and to rebuild the self-esteem of someone who feels utterly disillusioned by general practice at a later date. You should also bear in mind that most trainers want you to be happy in your placement and will do whatever is required to help you enjoy your learning experience. For more information on what to do if you aren't enjoying your general practice placement see **CHAPTER 15, 'LIFE EVENTS AND CHANGES THAT MAY AFFECT YOUR TRAINING'**.

But, in the hope of getting you started out on the right foot, here are my all-important two checklists which summarize what you should hope to find out in the first few weeks! Enjoy and good luck!

Ideally at the end of your first day you should know:

1. What to do in a fire
2. Where the resus trolley or bag is
3. How to summon help in an emergency
4. What your expected hours of work are
5. What your induction timetable involves
6. What your computer passwords are
7. When you will be taught how to use the computer and by whom
8. Whether or not you will have your own room
9. Where the toilet and coffee machine are
10. When your first meeting is with your trainer/educational supervisor

Ideally at your first meeting with your educational supervisor you should discuss:

1. The surgery's preferred way of documenting clinical information in patients' notes
2. Where you will find your doctor's bag

3. When you will start seeing patients—including length of appointments, catch-up breaks, and number of patients seen
4. Tutorial times—including timing, length, and proposed content
5. How to access your ePortfolio
6. What is expected from you in terms of analysing bloods, filing post, and prescribing (including where you can find a copy of the practice formulary)
7. When you will start doing home visits and how they will be assigned
8. When or if you will be included on the on-call rota and what that involves
9. How they would like you to approach referrals (e.g. would they like you to discuss them all first?)
10. Who you can go to for help and how it is best to approach them (e.g. phone call mid-surgery versus waiting outside door, etc.)

REFERENCES

1 Simon C, Everitt H, and van Dorp F (2010). *Oxford Handbook of General Practice*, 3rd edn. Oxford University Press, Oxford.

2 BMA. *Employment and contracts*. Available at http://www.bma.org.uk/employmentandcontracts/employmentcontracts/junior_doctors/framecontractGPregs0707.jsp (accessed 13 Nov 2009).

3 GPnotebook. Available at: http://www.gpnotebook.co.uk (accessed 14 August 2010).

CHAPTER 12

Stop and check! Are you ready for your ARCP panel review?

By now you should be very familiar with the different aspects of the training scheme along with the associated assessment process and are probably well on your way to getting your ePortfolio looking healthy. However, if your ePortfolio has started to metaphorically 'gather dust' over the last few months, then make sure that you do something about it before you are called up in front of the ARCP panel to explain your lack of house work. As discussed in earlier chapters, it really isn't appropriate to fill your ePortfolio with lots of irrelevant log entries the night before your educational supervisor's review, just to make it look like you have done something. Not only will your educational supervisor be unable to read and comment on these entries at such short notice, but it will be very obvious to the ARCP panel if you have 'padded out' your ePortfolio at the last minute. This chapter aims to illustrate some of the more common reasons for unsatisfactory ARCP outcomes in an attempt to motivate you to make your ePortfolio something that you can be proud of.

ARCP checklist

1. **Learning log** Make sure that you have a sufficient number of entries in your learning log and that they are reflective (e.g. they explain what you really learnt from the experience and how it will affect your future practice). You should be aiming to submit around two good quality entries per week and should avoid linking more than two curriculum headings

to each entry unless there is strong evidence within the written component of the entry to suggest that more headings are appropriate. It is also important to try and cross reference your learning log to your PDP.

2. **Skills log** Check that you are making progress with your skills log and aim to complete those DOPS which are relevant to your current postings. Although it isn't until ST3 that the ARCP panel will be checking that you have completed your mandatory DOPS, organized trainees will get them sorted and signed off as soon as possible. An important point to labour is that rating yourself as competent in the mandatory skills is not a substitute for the DOPS being independently verified by an authorized assessor (e.g. it is not acceptable to get your friends to sign them off for you!).
3. **Educational supervisor's review** Make sure that you have adequately prepared for your review and that you have given your supervisor sufficient time to evaluate the entries that you have submitted. You must make sure that you have completed the self-rating of competencies prior to your review as failure to do this is a common reason for an 'unsatisfactory' ARCP outcome. In addition, if your review has not taken place in enough time prior to the ARCP panel date then you may automatically be deemed 'unsatisfactory', as your review will have 'missed the boat' for the pre-panel evaluation of your ePortfolio. For standard August to August trainees, you should be making sure that your educational supervisor's reviews are done in November and May (i.e. completed in plenty of time prior to the panel dates). For more information on how to prepare for your review see **CHAPTER 6, 'SUPERVISORS, TRAINERS, AND HOW TO MAKE THE MOST OF YOUR MEETINGS'.**
4. **PDPs** Poor PDPs are another cause for unsatisfactory outcomes and so it is essential that you make sure not only that yours is up to date, but that the entries are SMART (Specific, Measurable, Achievable, Realistic/Relevant, and Timely). Whilst this sounds like a bit of a cliché, checking that you tick all of these boxes will help to make your PDP entries of good quality.
5. **Out of hours (OOH)** Make sure that you clearly document in your ePortfolio the number of OOH sessions worked including the amount of hours per session. Remember that you need evidence of at least 72 hours, so don't forget to either keep a list or enter them each time you work a session. Just saying that you have worked 12 shifts is not enough unless you clearly state the length of each of these and/or scan in OOH attendance forms.
6. **Absences** If you have taken any time out from your training for sickness or maternity reasons, then you must ensure that this is evident within your ePortfolio. The panel will not look kindly on those trainees with unexplained or unauthorized absences, therefore if you have been absent for genuine reasons then it is your responsibility to discuss this with your educational supervisor and make sure that the reasons are obvious within your ePortfolio.

7. **Mitigating circumstances** If there really is a genuine reason why you have not managed to complete the relevant components of your ePortfolio to the required standard then you must fill in the associated 'mitigating circumstances' form and make sure that the panel are aware of this. Whilst it is unlikely to prevent you receiving the 'unsatisfactory' verdict in your review, it may influence the support that you get from the deanery (and prevent you being 'released' from the training scheme!).

How will I know if an ARCP review has taken place?

This should be obvious when you first login to your ePortfolio homepage where you will find a link that says 'You have 1 unsigned Deanery Panel Review click here to accept'. Clicking on this link will enable you to view your ARCP 'verdict', which if found to be 'unsatisfactory' will give the reasons for this verdict as well as details of any remedial action which must subsequently take place. Regardless of the verdict, you will need to accept the review which can be done by clicking the 'Trainee signoff' box at the bottom of the page. Once you have accepted the review you will be able to view and save the ARCP review as a PDF file via the Completed 'Progress to Certification' link on the left toolbar. I would suggest printing out a copy of this and keeping it in your revalidation file for future reference.

Summary

It really shouldn't be difficult to get through the annual panel review if you are keeping your ePortfolio up to date and completing the necessary WPBA assessment tools at the correct rate. Being given an 'unsatisfactory' panel verdict is likely to lead to unnecessary stress and anguish and can be avoided by just keeping on top of things. If you know that you are struggling to get certain aspects completed and have a panel date looming, then make sure that you discuss this with your educational supervisor as soon as possible so that you can receive any help and support that you might need.

REFERENCE

1 RCGP. *The Quick Reference Guide to GP Training and Professional Development 2009*. Available at: http://www.rcgp-curriculum.org.uk/pdf/curr_Quick_Ref_Guide_to_GP_Training_and_Prof_Devt_mar09.pdf (accessed 12 August 2010).

CHAPTER 13

Thinking ahead and how to make yourself more employable at the end of it all

With over 3000 people joining GP speciality training each year, there is a common concern that come the end of it all, it may be very difficult for successful trainees to find a job in their chosen area of work. With this in mind, it's therefore worth thinking very early on about how you are going to make yourself stand out from the crowd, seem attractive to potential employers, and prevent your CV from being committed to the recycling bin as soon as it arrives in the post. Given that having the MRCGP is now compulsory for all new GPs, you need to try and see things from the employer's perspective and think about what additional skills and qualities you could bring to the practice team. During the ST1 and ST2 years you are less likely to have any MRCGP exams to worry about, so this may be a good time to get involved with some additional activities such as becoming your local BMA representative, completing diplomas and certificates, or attending some useful courses in order to improve your knowledge and skills in specialist areas. That's not to say, however, that you should sign yourself up for every course or diploma on offer. The GPST years are busy and stressful enough (and expensive!) without putting extra pressure on yourself to commit to things that either don't interest you or that you have no intention of following through to the bitter end. Instead, think early on about what aspects of general practice interest you most and where you see yourself heading in five to ten years' time. Most GPs have at least one specialist area of interest or additional commitment outside of ordinary general practice and working out what this could be for you will help you to get on the right tracks from the start.

This chapter aims to give you a brief introduction to some of the possible ways in which you can get some additional experience and qualifications outside of the world of ePortfolios and RCGP assessments, but is by no means an exhaustive list, as the possibilities and opportunities really are endless.

Diplomas and certificates*

Even though the new RCGP curriculum requires you to embrace the clinical specialities more than has been obvious in the past, there is still a lot to be said for making the most of your speciality posts and getting some definitive proof of what you have achieved during that time, in the form of diplomas or certificates. I don't mean simply because the piece of parchment paper that you get at the end of it all will look impressive on your consulting room wall, but because very often working towards an exam or assessment helps to focus the 'on the job' learning by making it seem more interesting, relevant, and real. If you are keen to achieve some additional qualifications during your speciality training then there are plenty of courses and diplomas to choose from, with more and more cropping up each year. Here are just a selection of the most popular ones with some information on what they involve and what they will add to your CV.

Diplomas

FAMILY PLANNING (DFSRH FORMERLY DFFP)

This diploma is extremely popular amongst GPSTs and not surprisingly, given that issues surrounding contraception and sexual health are extremely common amongst the patients that we see in general practice. There is also the additional motivator known as QOF, which back in 2009 introduced a new objective for trying to get GPs to increase the popularity of LARC (long-acting reversible contraception) methods, in an attempt to reduce the overall incidence of unwanted pregnancies and improve the cost-effectiveness of the contraceptive options on offer. Without your family planning training you will be unable to fit coils or contraceptive implants (e.g. LARC methods) and in fact may have very little knowledge on how they work and which patients would be suitable for what. You may also have very little experience of prescribing simple contraceptive options such as pills and barrier methods and therefore the DFSRH is likely to provide you with the all-essential knowledge base required for all types of contraceptive consultations. As well as completing the diploma, there is also the opportunity to gain what's known as the 'Letter of Competence' in both intrauterine techniques (LoC IUT) and subdermal implant insertion (LoC SDI). These qualifications allow you to fit coils and contraceptive implants unsupervised and are a mandatory requirement by many PCTs if practices wish to gain the necessary payments for carrying out these procedures. Ideally you should not be fitting these devices unless

*Fees and other costs are accurate as of 2010, but may since have changed.

you have the relevant faculty accreditation, although many GPs who gained these skills prior to the LoCs being introduced still continue to do so.

> 'Knowing from an early stage that I wanted to get my qualifications in family planning, I attended the theory course within the first few months of starting my VTS training. I then had to search far and wide to find a clinic that would accept for me for my practical training as places were few and far between with waiting lists as long as 12–18 months! At that point it was still called the DFFP (Diploma of the Faculty of Family Planning) and it was common knowledge that practical training placements were like gold dust, making it a fairly unique and impressive qualification to have.'
>
> ST3

> 'The DFSRH is an invaluable addition to your CV—I'm convinced that having obtained it was one of the reasons why I was successful in getting my first salaried post.'
>
> ST3

The DFFP became the DFSRH in 2007, after the college decided to rename the division to incorporate not only family planning but sexual health issues as well. As a result, dramatic changes were made to the way in which the DFSRH was taught and delivered and a completely new training process was introduced at the beginning of 2010. Gone are the days of the two-day DFSRH/STIF theory course followed by endless clinic sessions trying desperately to see 'x amount' of coils and 'y amount' of implants and instead the new sophisticated (if not perhaps a little complicated) scheme is now in full fledge.

'OLD STYLE' DFSRH (FORMERLY DFFP)

Those candidates who attended the DFFP theory course prior to December 2009 will still be eligible to continue with the old system of completing the practical training along with the logbook (which you should have been given at the theory course). You will need to contact one of the practical training programmes listed on the FSRH[1] website in order to be assigned a primary and secondary trainer for this. Throughout your practical training you will need to complete a minimum of four assessments with a faculty registered trainer. These are normally split into the following:

- Initial assessment with primary trainer
- Interim assessment with secondary trainer
- Interim assessment with primary trainer
- Final summative assessment with primary trainer

All of the experience that you gain during your practical training should be documented in your log book and once all of the competencies have been attained you should organize the final summative assessment with your primary trainer. It is **your** responsibility to organize your assessments and make sure that you have achieved the required learning objectives and competencies. Once you have successfully completed the DFSRH logbook, your primary trainer will be able to complete Part C which certifies to the FSRH that the diploma can be awarded. Your completed logbook plus the required registration and subscription fee should then be posted to the FSRH who will approve your application on the last working day of the month in which it is received. The registration fee is fixed at £60 but the subscription fee depends on when during the year you apply for your qualification and on average works out about £20 per quarter. You will also need to pay the relevant training programme for the pleasure of completing your practical training, with costs ranging from around £250 to £500.

My theory course was ages ago—do I need to repeat it before I can do the practical training?
The faculty states that your practical training should be completed within three years of attending the theory course; however, it is possible for you to apply for an extension of this time frame if you know of a reason why there may be a significant delay. In these circumstances, your primary trainer will normally recommend that you complete the online e-learning modules which form part of the new DFSRH (see below) in order to refresh your memory before you start the practical training. If you need to apply for this extension, then the necessary forms can be downloaded from the FSRH[1] website.

How do I gain the additional LoCs through the old system?
If you would also like to complete the LoCs for subdermal implant insertion (LoC SDI) and intrauterine techniques (LoC IUT) then you can start to do this during your DFSRH practical training. LoC SDI can be done alongside your practical training and requires you to record all insertions and removals in your logbook. These, however, must be observed by a trainer who holds the LoC SDI qualification and not just a GP who has been fitting them for years without being registered. LoC IUT can also be started during your practical training but cannot be completed until you have gained your DFSRH. Up to five IUD insertions completed during your practical training can be used for the LoC IUT but the remainder must be done once you have successfully gained your diploma. Your primary trainer should be able to give you more information on how best to achieve this. Each LoC costs £35 (by this I mean to get the certificate—the actual training is likely to cost you around £250) and requires you to keep up to date with your annual FSRH subscription.

THE NEW DFSRH

Those who have not attended the old-style theory course but would like to consider completing the DFSRH will now follow the new training process which consists of six stages.

- Stage 1—read the training requirements for the DFSRH as found on the FSRH website[2]
- Stage 2—register with the FSRH. This is achieved by completing the registration form[3] and returning it to the Faculty with the required payment

of £50. Once registered you will be allocated a FSRH ePortfolio which allows you to record your progress throughout the various components of the diploma

- Stage 3—complete the **e-SRH** online e-learning modules (see later question **'WHAT ARE THE ONLINE E-LEARNING MODULES?'**)
- Stage 4—complete a '**Course of 5**' (see later question **'WHAT IS THE "COURSE OF 5"?'**)
- Stage 5—apply to a general training programme to complete the **Clinical Experience and Assessment** component (see later question **'WHAT IS THE CLINICAL EXPERIENCE AND ASSESSMENT COMPONENT?'**)
- Stage 6—apply for the Diploma using Form H (this must be done within 12 months of completing your DFSRH training programme). You will need to submit the following:
  - *ePortfolio final assessment*
  - *Evaluation of DFSRH Clinical Experience and Assessment*
  - *Appropriate fee*

What are the online e-learning modules?

These are found within the e-Learning for Healthcare (e-LfH)[4] website under the project heading 'Sexual and Reproductive Healthcare e-learning' (e-SRH). You may already be aware of some of the 40 or more other e-LfH online modules such as e-GP and the Foundation e-learning programme, which are also delivered via this website. Access can be gained via the home page by clicking on the left-hand tool bar link entitled 'sign up for access'. Once you have been allocated your dedicated user name and password, you are then free to search for and utilize all of the different projects available.

The e-SRH replaces the 'old-style' theory course and requires you to work your way through the online modules and then complete a final summative assessment at the end. You can stop and restart the modules at any time, completing them entirely at your own pace. Once you have been successful in completing all components of the e-SRH, including the final exit assessment (module 15), you will then be eligible to attend a 'Course of 5' which is the next step in attaining the diploma.

What is the 'Course of 5'?

The 'Course of 5' is a series of five one-hour sessions with fixed content and specific assessments, which aim to put some of the theory learnt during the e-SRH into practice. The five sessions will normally be completed by attending a one-day 'Course of 5' study day but in some areas may be delivered in different ways. Before attending a 'Course of 5' you must have successfully completed the e-SRH modules and have also had the pre-course requirements signed off by your educational supervisor within your FSRH ePortfolio (not to be confused with your GPST ePortfolio!). This demonstrates to the Faculty that you have the baseline gynaecological competencies required to continue with the diploma. For GPSTs this should not be a huge issue as you should hopefully have already achieved a DOPS in vaginal and speculum examination during either an O&G or general practice post. You will need to provide evidence of this along with your e-SRH completion certificate before you will be allowed to attend the 'Course of 5' study day. This should simply involve printing out the relevant section within your FSRH ePortfolio.

Each session within the 'Course of 5' has a fixed content and requires active participation by all candidates. The maximum ratio between candidates and facilitators on these courses is four to one, to enable all candidates to get involved and get the maximum benefit from the interactive sessions.

The session titles are as follows:

- Session 1: Taking a sexual history and HIV pre-test discussion and testing
- Session 2: STI screening and testing and teaching the use of condoms
- Session 3: Practical aspects of contraception
- Session 4: Young people—consent, confidentiality, Fraser guidelines and safeguarding children
- Session 5: Managing sensitive scenarios

Across these sessions there are a total of nine assessments, at least seven of which must be successfully completed by candidates by the end of the 'Course of 5'. It therefore certainly isn't one of those study days where you can fall asleep at the back of the room and get through the whole day without saying a word. As mentioned previously, it requires all candidates to get actively involved in order for them to successfully get through this part of the diploma. The details of where you can attend a 'Course of 5' can be found on the FSRH[5] website, although in many areas of the UK they are still in the process of being developed, so may not appear on the website until late autumn 2010.

What is the Clinical Experience and Assessment component?

This replaces the old 'practical' element of the DFSRH and is intended to regulate and simplify the way in which the practical training is delivered. Once you have completed your 'Course of 5' you will need to contact your local General Training Programme[6] in order to be assigned a faculty registered trainer. At the first meeting you will review your ePortfolio entries and assess which competencies you still have yet to achieve. Make sure that during your regular ST training you are not only submitting interesting sexual health entries to your ST ePortfolio, but that you are 'copy and pasting' them to your FSRH ePortfolio as well. The reason for this is that part of the Clinical Experience component requires the primary trainer to look at the consultations documented within your FSRH ePortfolio and assess your competence in the following seven topic areas:

1. An effective contraception choices consultation
2. Consultation for a woman wishing to use an oral or injectable contraceptive, patch, or vaginal ring
3. Assessing and advising a woman wishing to use an intrauterine method or subdermal implant, prior to insertion

4 Responding to a request for emergency contraception
5 Taking an appropriate history and assessment of a woman with bleeding problems whilst using hormonal method
6 Taking an appropriate sexual history and risk assessment for STI and pregnancy and performing the appropriate tests for an asymptomatic woman or man requesting sexual health screening
7 Taking an appropriate history and assessment of a woman with vaginal discharge or pelvic pain

Each topic requires at least one assessment in the form of either an ACP or a RDCP. Just when you thought you'd heard enough acronyms…

ACP stands for Assessment of Clinical Practice and is very similar to a COT in that you are observed in a consultation with a service user. RDCP stands for Reflection and Discussion of Clinical Practice and is similar to a CbD in that you discuss a consultation that you have had with a service user with your primary trainer, without it having to have been observed by them. You are required to do a minimum of four ACP assessments with three of them being done in topic areas 1, 3, and 6 or 7. Sound confusing? I agree, but in real-life terms it's simpler than it sounds. Once you have achieved a successful assessment in each of the seven topic areas, it means that you have been judged as being competent at the level of independent practice in these areas.

As well as achieving competence in the required seven topic areas (via successful completion of the required ACPs and RDCPs), you are also required to obtain six consultation feedback forms from sexual health consultations that you have led. It is recommended that at least some of these feedback forms should come from the consultations that you have used for your ACPs and RDCPs, as this will enable both you and your trainer to get an idea of how you come across to the patients. Following this, you will be able to organize your final assessment with your trainer, during which you will review your consultation feedback forms and discuss any outstanding issues that may have become apparent during your training. Should everything be completed successfully then you will be able to apply for your diploma as described in Stage 6 of the training process.

There is no suggested time frame in which the Clinical Experience element should be achieved, although you must complete the diploma within three years of your exit assessment from the e-SRH. It's also worth noting that you must recertify for your diploma every five years and that in order to continue using the qualification you must pay the annual subscription fees to the Faculty.

How do I gain the additional LoC for the new DFSRH?

For those trainees following the new training process you will no longer be required to watch CD ROMs or attend a model arm training course for the LoCs, as these areas will already have been covered within modules 17 and 18 of the e-SRH and the 'Course of 5'. You can therefore complete the practical training for the LoCs at the same time as your DFSRH Clinical Experience, although they cannot be awarded until after you have successfully gained your diploma. At present you will still need to apply using the paper documents which can be accessed via the faculty website.

Box 13.1: Cost of obtaining the new DFSRH

e-SRH:	Free
Course of 5:	£250–£400
Clinical:	£250–£400
Registration fee:	£60
Subscription fee:	£20–£80 per year
Minimum total:	**£580**

DERMATOLOGY

There are several different dermatology diplomas available across the country, but the one which has been accredited by the RCGP Higher Professional Development panel is the Diploma in Practical Dermatology delivered by Cardiff University. This is the diploma which you will require should you wish to become a GPwSI or clinical assistant in dermatology, but is also a worthwhile consideration if you just wish to increase your knowledge and skill in this area for either the benefit of yourself or your surgery.

The diploma involves a one-year postgraduate distance learning course, which is primarily aimed at GPs who wish to gain expertise in the practical management of skin disease. It is predominantly delivered online and requires you to complete a number of components:

- Four pieces of written work of 1500 words each—this can take the form of either a case-based scenario, a new practice guideline, a patient hand-out, or an essay style question
- Interactive online tutorials with a dermatologist
- Writing up case histories
- Participating in online forums
- One online group assignment each semester

Following this you will be required to sit the final exam which is held in June each year in either Cardiff or Hong Kong. Of note, the qualification is recognized not just in the UK but also in Hong Kong, Singapore, and Australia, for those who have plans to move overseas. It is suggested that you will need to commit around 10 hours per week

Box 13.2: Minimum cost of obtaining the Diploma in Practical Dermatology

Total:	**£3,495**

of your time in order to successfully complete the diploma and the course runs from September to June each year. Places are offered on a first come first served basis and the cost for 2010 was £3,495, which can be paid in three instalments of £1,165 each.

> 'I completed the Cardiff University Diploma in Practical Dermatology during my ST3 year. I really enjoyed the diploma but it was a lot more work than I'd anticipated, especially when trying to prepare for MRCGP examinations and doing OOH sessions as well. If I was recommending this to another trainee, I would advise doing it in ST1 or ST2 so you can get full enjoyment out of it.'
>
> ST3

OBSTETRICS AND GYNAECOLOGY (DRCOG)

The DRCOG is issued by the Royal College of Obstetricians and Gynaecologists and exists to recognize the appropriate knowledge of women's health care as applied to a GP in the UK. It consists of two written papers that are taken on the same day. Examinations are held in various centres across the UK biannually in April and October. There is no specific eligibility criteria in terms of length of time spent working in the speciality of O&G before applying for the exam and in fact you can still have a go even if you have never done any O&G as part of your training.

The exam lasts three hours in total and is split into two separate papers with a 15-minute break in between. Paper 1 lasts one and a half hours and consists of 30 EMQs plus 18 'best of 5' questions. Paper 2 is also one and a half hours long but contains 200 multiple choice 'true/false' questions. You are given a cumulative mark for both papers and therefore do not have to pass each paper individually. Instead it is your overall score which determines whether or not you pass or fail. Should you fail the exam, you are only allowed a further four attempts. As with most of the other diplomas you can access practice questions online via websites (such as http://www.onexamination.com and http://www.pastest.co.uk) and you may find that the undergraduate books that you used for your O&G placement at medical school may come in handy for brushing up on some of the anatomy and physiology questions. To sit the exam costs around £400 and obviously there are no discounts for any further attempts required. Studying for the DRCOG is not only beneficial for those working in an O&G post but can also be a useful revision tool for those trainees that are unfortunately not spending any time in O&G during their training. Sitting the diploma not only consolidates your previous knowledge of women's health but also puts it in the context of primary care, which can be desirable to potential employers.

Box 13.3: Minimum cost of obtaining the DRCOG

Exam costs:	£400
Total:	**£400**

'I decided to do DRCOG as I was doing a six-month placement in O&G. If you want to a get further qualification in this area, then DFSRH is much more important so I would only recommend DRCOG as an addition to this.'

ST3

CHILD HEALTH (DCH)

The Diploma in Child Health (DCH) is issued by the Royal College of Paediatrics and Child Health and comprises of a written paper plus a clinical exam. The written component is Paper 1a of the membership exams for those doctors carrying out paediatric specialty training and consists of 12 'EMQ's, 15 'true/false questions', and 48 'best of 5' questions. The syllabus along with child health surveillance guidance and other guidelines relating to vision, hearing tests, and language development in children, can be found on the DCH section of the website[7]. A specimen exam paper is also available in order to give you an idea of the standard and type of questions that you will be expected to answer. Online revision courses are available through companies such as onexamination.com and PasTest, although the RCPCH suggests that the best type of revision is to subscribe to the MPCPCH Mastercourse which consists of a DVD and online self-testing. This costs around £200 and is probably more beneficial for those committing to the full membership exams (which you shouldn't be); however, it's worth a look if you have the cash to spare.

The written exam can be sat three times a year in January, May, and September at the following centres: Belfast, Birmingham, Cardiff, Dublin, Edinburgh, Glasgow, Leeds, London, and Sheffield. The places at each centre are allocated on a first come first served basis so make sure that you book your place early in order to prevent you having to travel half way across the country to sit it. The written paper costs for 2010 are £214 and you must successfully complete this component before you are eligible to apply for the clinical exam. There is, however, no limit to the number of attempts which you can have to pass the written paper.

The clinical exam runs twice each year in March and November and is carried out at various centres across England and Wales. Unlike the written paper, you cannot choose your centre location although your home address is taken into consideration where possible. The RCPCH recommends that you should have at least six months' experience in paediatrics before attempting the clinical exam although this isn't compulsory. Each attempt at the exam costs £365 and if you are unsuccessful on three occasions then you will be required to resit the written paper before you can resit the clinical exam for the fourth time.

Once you have successfully completed both the written and clinical aspects of the diploma you will be issued with the certificate, but not before you have paid the RCPCH £85 for the pleasure of them doing so.

Given that a large proportion of the patients that you see on a day-to-day basis within general practice will be children, this can be a very useful diploma to have. Having the confidence to manage children in the primary care setting is an invaluable asset and is one which is likely to be very appealing to prospective employers.

Box 13.4: Minimum cost of gaining the DCH

Written:	£214
Clinical:	£365
Diploma:	£85
Total:	**£664**

'I passed the DCH written paper, but failed the practical, and didn't sit it again. I did find that studying for the diploma helped me get the most out of my pediatrics placement, but long delays between the written and the practical, coupled with other commitments (came back from honeymoon and had practical and DRCOG within a few weeks) means it is tough. I think four-month jobs will make it even more difficult to balance. But... it encourages self-directed learning during the placement and I felt it was worth doing for that reason alone.'

ST3

GERIATRIC MEDICINE (DGM)

This diploma is issued via the Royal College of Physicians and exists to recognize competence in the provision of care to the elderly. It is aimed primarily at GPSTs who wish to either specialize in this field or simply improve their knowledge of elderly care in preparation for the UK's aging population. Part one is held at the Royal College of Physicians and consists of a two-and-a-half-hour written exam containing 60 'best of 5'questions. It is held twice a year in February and July and costs £200. On successful completion of part one you are then eligible to apply for the part two clinical exam, which is held in various centres across the country in May and November. This involves an OSCE style examination with four stations, which assess your history taking, communication, and clinical examination skills in the context of elderly care. This part of the diploma costs £295 and must be successfully completed within two years of passing part one. Unlike some of the other exams, there is no limit to the amount of attempts that you can have at passing. This diploma may be particularly useful if you are hoping to join a practice that has a large proportion of elderly patients within their demographic. Elderly care is a tricky speciality to master, given the issues of multiple

Box 13.5: Minimum cost of obtaining the DGM

Part one:	£200
Part two:	£295
Total:	**£495**

comorbidities and polypharmacy and if it's an area which you find challenging and enjoy, then this may be the diploma for you.

Certificates

SUBSTANCE MISUSE CERTIFICATE

This certificate was created by the RCGP Substance Misuse Unit (RCGP SMU) as a way of improving the care delivered to patients suffering from substance misuse. Part one of the certificate is aimed at GPs who provide advanced service level treatment as part of shared care or a locally enhanced service (LES). It is therefore worth considering if you plan to work within an area where drug misuse issues will form a significant proportion of your daily work load. Part one consists of two online e-learning modules which are accessed via the RCGP education section of doctors.net.uk[8] (and require you to log in to the website) plus a 'face to face' training event. Details of where you can attend the training events can be found on the RCGP website[9]. Whilst the e-learning modules are free to access, the training event costs anywhere between £150 and £250 depending on where you attend the course. If you don't feel that committing to the full part one certificate would necessarily be relevant to your area of practice, then it may still be worth completing the free online e-learning modules in order to improve your knowledge in this area, as it is often a part of the RCGP curriculum which is difficult to cover for some trainees.

Part two of the certificate is aimed at either practitioners with a special interest in working with substance misusers or individuals intending to take on a leadership or strategic role. It consists of three formal training days plus the equivalent of six self-directed learning days which are then assessed in a final viva with a tutor at the end of the course. The work is then reviewed by the RCGP SMU assessment board who decide whether the candidate should pass or fail. Each course runs from February to October and costs £1,500. It is unlikely that you will benefit from committing to part two of the certificate unless you have a special interest in this field but it may be something to think about in the future if you join a practice which may benefit from this level of expertise.

ALCOHOL CERTIFICATE

This was introduced by the RCGP in 2009 following the success of the Substance Misuse Certificate and the increasing demand for education in the treatment of alcohol misuse. Similar to the Substance Misuse Certificate it involves completing an e-learning course[10], followed by a one-day training event and completion of a self-directed workbook. The one-day training event varies in cost from £150–£250 depending on whether you attend a local or national event. Given that alcohol misuse is becoming an increasing problem in most practice populations, this is a certificate which is likely to be particularly useful, as again this is often an area of the curriculum which is both difficult and challenging to cover.

Courses

If you are not keen to pursue any of the diplomas or certificates but would like to expand your knowledge in other ways, then there are a whole host of courses offered by the RCGP and elsewhere, which are likely to be significantly cheaper and perhaps require less ongoing commitment than a diploma. Busting a gut trying to complete every diploma and attend every course on offer makes neither financial nor emotional sense since a frazzled overspent GP is no good to anyone! Instead you need to decide which additional qualifications or skills will suit you best and more importantly fit into your work–life balance (which is, after all, why we chose general practice in the first place isn't it?). Listed below are a few examples of some of the courses run by the RCGP; however, many more can be found via the 'Courses and Events' section of the RCGP website. Remember that for any of the courses held in London which require an overnight stay, you are eligible for discounted rates at the RCGP accommodation venues.

Minor surgery

This is usually held at the RCGP headquarters in London and lasts for two days. After a few lectures on the theory side of things, you are given the opportunity to practise cutting and sewing using proper instruments and model skin. You also get to practise the use of cryosurgery and can have a 'stab' at doing some joint injections on fake arms and legs. It costs around £400 for RCGP members but sells out fast so make sure you book early to avoid disappointment.

Health and work in general practice

This is a half-day course run in various locations across the country that looks at the way in which unemployment can have harmful effects on health and looks at ways in which those patients suffering with a chronic illness can try and get back to work. This course is likely to appeal to those who get frustrated with the endless amounts of sick notes that we are requested to write these days and is likely to be well worth the £15 that it costs to attend!

Enhancing skills in primary care ophthalmology

This is a one-day workshop held at the RCGP in London, which using the RCGP curriculum as a guide aims to expand your knowledge on common eye problems seen in primary care. It also addresses the issues relating to disability, screening, and compliance and how to coordinate with community optometrists and other specialists.

Representing bodies and getting on committees

There are several organizations on both a local and national level which require volunteers to get involved and represent their peers. Examples of these include the BMA, the RCGP, and your local deanery or LMC. If you are elected to represent other GP Specialty Registrars on any of these recognized bodies and are required to attend any conferences or meetings as part of this role, then you will be given appropriate special paid leave to undertake such functions. You should wherever possible inform your trainer or educational supervisor of any such commitments before you commence your attachment and must also obtain their consent prior to each absence from duty. This consent is unlikely to be withheld unless there are exceptional circumstances in the reasonable opinion of the trainer or educational supervisor, and if agreed, these absences should be considered as included in your working time commitment. If subsequently any of these commitments significantly impact upon your learning, then an extension to training may be required at the discretion of the Director of Postgraduate GP Education.

Other stuff

If none of the above appeals to you or it seems as though everyone has these extra diplomas or certificates, then try to think of novel ways in which you can expand your knowledge and make your CV stand out. These may include carrying out research projects or audits, putting on a GPVTS medics' revue, or organizing a community project for a local charity or organization (which may have the additional benefit of enabling you to sign off some of the more difficult competency statements!)

REFERENCES

1 Faculty of Sexual and Reproductive Health. Available at:http://www.ffprhc.org.uk (accessed 14 August 2010).

2 Faculty of Sexual and Reproductive Health. *DFSRH Overview*. Available at: http://www.ffprhc.org.uk/pdfs/DFSRHOverview.pdf (accessed 14 August 2010).

3 Faculty of Sexual and Reproductive Health. *Registration form*. Available at: http://www.ffprhc.org.uk/pdfs/DFSRHTraineePreRegForm.pdf (accessed 14 August 2010).

4 eLearning for Healthcare. Available at: http://www.e-lfh.org.uk (accessed 14 August 2010).

5 Faculty of Sexual and Reproductive Health. *'Course of 5' contact details*. Available at: http://www.ffprhc.org.uk/pdfs/Course5List.pdf (accessed 14 August 2010).

6 Faculty of Sexual and Reproductive Health. *General Training Programme details*. Available at: http://www.ffprhc.org.uk/Default.asp?Section=ExamsTraining&Subsection=practical (accessed 14 August 2010).

7 Royal College of Paediatrics and Child Health. *Diploma in Child Health*. Available at: http://www.rcpch.ac.uk/Examinations/DCH (accessed 14 August 2010).

8 Doctors.net.uk. Available at: http://www.doctors.net.uk (accessed 14 August 2010).

9 RCGP Substance Misuse Unit. Available at: http://www.rcgp.org.uk/practising_as_a_gp/substance_misuse.aspx (accessed 14 August 2010).

10 Alcohol Learning Centre. Available at: http://www.alcohollearningcentre.org.uk/eLearning/IBA/ (accessed 14 August 2010).

CHAPTER 14

Thinking about applying for AKT? Some handy hints and tips…

Introduction

In exam terms I would say that the AKT is the first big hurdle that you have to face for the MRCGP. Whilst you should have been having regular formative assessments with your supervisors up until now, this is the first official, 'sit down in a dark room for three hours type of test'. And it is a test, in the true school-like sense of the word. As the acronym states, the AKT is designed to test your ability to apply the knowledge that you have learnt so far in a variety of different situations. It is NOT a test of how many different pages of the *BNF* you have memorized or how many NICE guideline flow charts you have etched on your bedroom wall. It covers general principles over a wide range of topics and whilst it doesn't expect you to have the knowledge of a consultant in each speciality field, it does expect you to have spent some time in general practice practising evidence-based medicine. Some of the following information can be found on the RCGP website, but hopefully this chapter provides more than just the factual information and instead gives a first-hand experience of what it really is like. Plus, having it all condensed in a book means that you can read it at your leisure, whether that be in the bathroom, on the train, or whilst waiting for the dentist.

General information

AKT—what does it stand for?

Applied Knowledge Test—it does exactly what it says on the tin.
The college definition, however, is a 'summative assessment of the knowledge base that underpins independent general practice in the United Kingdom within the context of the National Health Service'[1].

What does it involve?

The AKT is a computer-based exam consisting of 200 questions made up of a variety of different question types, e.g. EMQs (Extended Matching Questions), SBAQs (Single Best Answer Questions), tables and algorithms for completion, and data interpretation questions. Approximately 80% of question items will be on clinical medicine, 10% on critical appraisal and evidence-based clinical practice, and 10% on health informatics and administrative issues[1]. You will be required to apply the knowledge that you have learnt in the context of different clinical scenarios, rather than just regurgitating facts that you have memorized. Each correct answer gains you one mark and there is no negative marking.

How long does the exam last?

The exam lasts three hours and there is no break. You will be provided with your own work station that consists of a computer and a wipe-clean eraser board for making notes. The eraser boards must all be handed back at the end of the exam.

When CAN I sit the exam?

There are three AKT sittings each year—normally in January, May, and October. The college opens to applications about two months before the exam date and there is a relatively small window for applying and registering with Pearson VUE to choose an exam centre. The later you leave it, the less choice you will have in terms of location and morning versus afternoon sittings. All of the dates are subject to change so it's important to keep your eye on the RCGP website for further information.

When SHOULD I sit the exam?

The AKT can only be taken during the ST2 stage of specialist training or later unless you commenced specialist training on or before 1 August 2009, in which case you will be permitted to take the AKT during ST1 pending transition to ST2.Whilst it may seem tempting to try and get the AKT 'out of the way' at an early stage, bear in mind that the test is aimed at the level of a GP Registrar, who is deemed to have sufficient knowledge in order to practice independently. You will only 'get it out of the way' (as it were) if you pass. If you don't pass, not only have you lost a substantial amount of money but it is all recorded on your ePortfolio and there is no going back. It may also be worth considering that it won't be long before the college bring in the opportunity to more than just 'pass' these exams with the possibility of being awarded merits or distinctions with your qualification—all of which could be very important when it comes to the job race later on. No one wins any prizes for sitting it early, so you might as well wait until you are properly ready and prepared to pass with flying colours!

Where can I sit the exam?

Many of you may have already encountered one of the Pearson VUE centres where the AKT exams are held, when you sat your driving theory test. There are loads of them scattered across the country (more than 150 in fact) and the chances are that you won't have to travel too far to sit the exam. For example, within 30 miles of Birmingham there are 20 different centres to choose from, and which one you get depends on how organized you are when booking the exam.

How much does it cost?

At the time of this book going to press the exam fee was £390 but this is likely to increase as the years go by.

What's the pass mark?

This varies with each exam depending on how difficult it was deemed to be. It is not one of those exams where a certain proportion will definitely fail and instead the pass mark is decided by a panel of examiners. It averages around 65–75% although the April 2009 pass mark was only 63.3%[2]. The trend shows that those sitting the exam in ST3 tend to be more successful than those sitting it in ST2, so bear this in mind when planning when you would like to sit the exam.

Applying for the exam

I think I'm going to go for it—how do I actually apply?

Application is easy but you will first need to have registered for the MRCGP through the RCGP website. Fig. 14.1 shows the sequence of events that you should follow.

Once you've made the decision to go for it, book early so that you don't end up having to travel halfway across the country to sit the exam. The College do have reserve dates in case all of the spaces are filled but these will only be released if all centres are completely full on the day of the exam, so don't count on it!

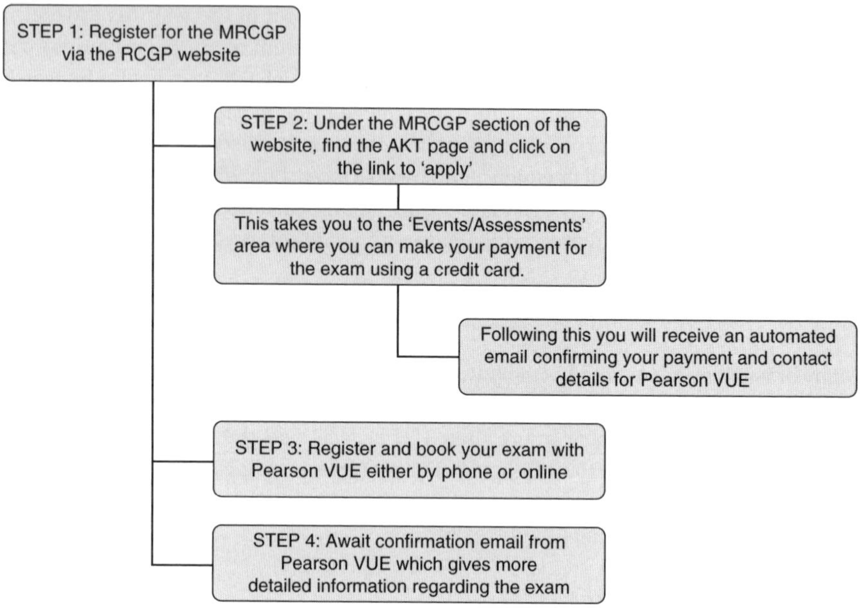

Fig. 14.1 Sequence of events when booking the AKT

Should I choose a morning or an afternoon session?

You are normally given a choice as to whether you would like to sit the morning or the afternoon exam when booking. Both will be exactly the same but if you take the morning exam you will be kept in quarantine until the afternoon candidates have been checked in and hidden in a separate room to avoid any cheating. Much of your choice will depend on the distance you have to travel to the venue and whether you're a morning person or not. Those types who stay up revising until the early hours the night before may be better off doing the afternoon session to give their brains time to recharge. Also, if you're a candidate that has been granted extra time for whatever reason you will need to book an afternoon session.

I've got a disability—will this affect my ability to do the exam?

No, it shouldn't do, but you must register this at the time of applying for the exam and request any reasonable adjustments that may be required, e.g. extra time.

Preparation for the exam

Should I sit the AKT before the CSA?

Many people ask whether they should sit the exam before the CSA and my immediate answer would be 'yes'. The reason being that sitting the AKT helps you to consolidate your learning and get you focused on what knowledge base you require to be a competent GP. The CSA is a totally different exam but it's much easier to concentrate on the practical elements once you have the basic knowledge base firmly sorted. It's definitely important to think about the timings of both exams from an early stage in your training, as certain life events may well contribute towards your decision of when to apply. For example, sitting your AKT the week before you get married will not only mean that the weeks running up to the wedding would be even more stressful, but you may well struggle to enjoy your wedding day and honeymoon if you spend the whole time worrying about the results.

> 'I chose to sit my CSA towards the end of ST2 as the first sitting in my ST3 year was scheduled for the week before I was getting married and I knew that all I would be concentrating on that time was napkin colours and wrapping favours and not NICE guidelines or statistics!'
>
> ST2

> 'I sat the AKT and CSA around the same time as I felt that much of the knowledge I was brushing up on for my AKT would help me in terms of diagnosis and clinical management in the CSA. It was very stressful but worth it as I managed to pass both first time.'
>
> ST3

> 'I stupidly left my AKT until the last minute as I didn't really feel that I was prepared for it any earlier. Luckily I passed first time but if I hadn't it might have meant needing to extend my training which would have been a disaster.'
>
> ST3

Do I have to pass AKT before I can sit CSA?

No, you don't, but it sometimes helps to have established your knowledge base before approaching CSA. See the previous question.

When do I need to start revising?

This totally depends on the type of person you are and how good your knowledge base is to start off with. Those who start revising too early often 'burn out' as the exam approaches and those who start revising too late may feel that there are topics and subjects that they don't have enough time to cover in enough depth. This is a very personal choice but three months before is probably a good average. Much of it will depend on what methods you use to revise and what job you're doing at the time. Working long, unsociable hours in A&E for example may not leave you with much book revision time, but will provide you with continuous 'on the job' learning in a variety of different specialities.

What do I need to know?

Contrary to popular belief you do not need to know ALL of the published NICE guidelines inside out or have memorized huge chunks of the *BNF* (although it's unlikely to do you any harm if you do enjoy doing that sort of thing!). What you do need however is a broad knowledge base that covers all of the common conditions listed in the curriculum document, as well as the business, administration, statistical, and legal aspects that are mentioned. Making sure that you have read the most recent NICE guidelines and are aware of their recommendations will definitely help and although no one knows the DVLA guidelines for every condition off by heart, it's important to have some knowledge of the more commonly used guidelines for driving, e.g. in patients with epilepsy. Learning the entire contents of a statistics book is also likely to be unhelpful, not to mention time-consuming, but understanding some of the more common principles such as false negatives/positives and odds ratios is quite important. The June 2009 issue of *InnovAiT* has a very useful article entitled 'An introduction to basic statistics in primary care'[3] which is well worth a read and gives you a brief rundown of some of the more important statistical principles which you need to understand. You should also have some grasp on ethical issues including consent and capacity and should have a basic idea of how a GP surgery runs and the role of audit and critical appraisal within general practice.

All of the questions are written by practising UK GPs and so you should not encounter anything too ambiguous or irrelevant to UK general practice. Remember, however, that you will be expected to answer the questions according to current UK national guidelines and not according to any local policies or practices. It's worth keeping abreast of some of the more recent and common NICE/SIGN guidelines, as these often form the basis of some of the questions within the exam, in order to assess that your knowledge of evidence-based medicine is up to date.

What's the best type of preparation for the exam?

This is a common question asked by many trainees and in actual fact the answers given vary hugely from one individual to another. There are many different ways in which you can prepare for the AKT, the most popular of which are listed below:

- *InnovAiT* journal articles and AKT questions[4]
- Online revision questions

- Forming study groups
- Sample AKT questions from RCGP website
- Pearson VUE tutorial
- Revision books
- Revision courses

InnovAiT journal articles and AKT questions

Within each month's *InnovAiT* there are AKT questions based on the topics covered in the edition, which have been written by RCGP-approved GPs and extensively peer reviewed. Using these articles and AKT questions as revision could provide the basis to a study group session or self-directed revision task. If you're someone that commits your *InnovAiT* journal to the recycle bin as soon as it's read, then you can access all of the past articles and the bank of AKT questions online once you have registered with *InnovAiT* which is free to all AiTs.

Also, look out for a new book of AKT questions that is currently being produced by *InnovAiT* and its contributors, which hopes to provide a large bank of peer reviewed practice questions for you to attempt prior to the exam.

> 'I used *InnovAiT* to help with my AKT preparation – it's college produced so most likely to be the right level/depth. Think of it as RCGP revision notes.'
>
> ST3

Online revision questions

The most common form of preparation is practising online exam questions which not only enables you to get used to the format of the questions, but also gives you some experience of what it's like to sit and work at a computer for three solid hours. It's harder than it sounds and for those who have only sat written exams in the past it really is quite different. There are lots of different websites which are providing online revision questions for a fee (some of them used to be free but they have obviously cottoned on to the market out there now!) and they vary in style and standard. The best way to get an idea of the standard of the questions that you will encounter in the exam is to have a go at the sample questions posted on the RCGP website. They can be found on the 'AKT' page under the MRCGP section of the website.

> Tip: The best way to get an idea of the standard of the questions that you will encounter in the exam is to have a go at the sample questions posted on the RCGP website.

If you do decide that you'd like to pay for some online revision questions, then there are several main websites to choose from.

ONEXAMINATION.COM

This is the most expensive of the four but probably the most commonly used and the one with the largest number of practice questions. It currently has a bank of 3150 questions which contain a mixture of SBAs, EMQs, and MCQs, although the website claims that they are adding 100–200 new questions per month. As well as providing a bank of questions it also gives you a detailed performance analysis and feedback on how you are doing in relation to those sitting the same questions across the country. Some candidates have said, however, that in retrospect the questions in this bank are perhaps more difficult than those encountered in the real exam so bear this in mind when making your decision and try not to get too downhearted if you aren't getting above 80% on all of your practice papers!

Prices vary from £48.94 for one month's subscription to £88.09 for four months' subscription although they do offer a 33% discount to BMA members. They also have a 'difference deal' policy which states that if you fail they will refund the difference between the price you paid and the cheapest competitor price (only valid when you purchase a three-month course priced at £78.30, giving you a refund of £58.30).

PASTEST.CO.UK

This is one of the other well-recognized online exam question banks which at present has around 1500 questions in its bank which are appropriate for AKT practice. It also enables you to monitor your progress and performance, allowing you to compare yourself with your peers and identifies areas of weakness in which you could improve.

A three-month subscription costs £33, compared with a six-month subscription which costs £53.

NPEP

This is a new online educational self-assessment tool produced by the RCGP in Scotland and aimed specifically at GP Registrars preparing for the AKT assessment. It includes 150 different AKT-style questions which are mapped to areas of the RCGP curriculum. In order to access this you will need to register for nPEP as an AiT—see http://www.npep.org.uk/.

PASSMEDICINE.COM

When I sat AKT three years ago, this website was free to access but at that time only had a very small bank of questions. Since this time it has fully developed and they are now charging for access to the questions although it is considerably cheaper than onExamination.com or PasTest.co.uk. It currently contains a bank of around 1500 questions and like the other websites allows you to compare your performance against others and calculate your average scores, giving you feedback on your performance. It also helps you to make revision notes based on each question and collates key points to help with last minute revision.

A four-month subscription costs only £20 and a six-month subscription is slightly more at £30. Many trainees have commented that the questions on this website are perhaps too simple compared to the real exam so bear this in mind when making your choice.

AKTREVISION.COM

This website holds around 1150 questions and also gives an analysis of your performance. A three month subscription costs £20.

> 'When revising for AKT, make sure that you have broadly covered the curriculum, as this way there shouldn't be any subjects that surprise you when they come up in the exam.'
>
> ST3

> 'Non-college-produced online revision questions can be a useful when guide revising for AKT, but individual sites are not always pitched correctly (some are a bit more MRCP than MRCGP!) So take some of them with a pinch of salt and read around.'
>
> ST3

FORMING STUDY GROUPS

Getting together in small groups to tackle your revision for AKT is a good way of covering lots of material in a short space of time. Meeting once a week to discuss specific topics that you have decided on beforehand, means that not only are you sharing the work load but it also allows you to learn from others who may have more knowledge and experience of that chosen subject than you.

SAMPLE AKT QUESTIONS FROM RCGP WEBSITE

As mentioned previously, the RCGP website does have a page of sample AKT questions that you can print off and attempt, in order to give you an idea of the standard of questions that you will be asked in the exam. However, it does not give you the correct answers to these questions but discussing which answers are right and wrong could form the basis of one of your study group sessions.

PEARSON VUE TUTORIAL

This should form an essential part of your AKT preparation, as it allows you to become familiar with the way in which the exam will be laid out on the computer. It's an online-based tutorial which can be found on the Pearson VUE website (http://www.pearsonvue.co.uk) and shows you the different formats in which the questions will be presented and the ways in which you choose and edit your answers. You will also have the opportunity to do a few practice questions at the start of the exam (which are not assessed) but it's definitely worth doing the Pearson VUE tutorial in advance of this.

REVISION BOOKS

Since the birth of AKT, there have also been many new books of AKT questions published by individual authors and companies. These all vary in price, standard, and quality and can be useful if you don't have easy access to the Internet or spend a large amount of time on public transport. There are also many books around that were written for the old-style MRCGP MCQ exam, which you may find useful although be aware that a large amount of the evidence-based medicine questions will be out of date.

If you'd like to read a book that might shed a bit more light on the statistics elements of the exam then PasTest have produced a book entitled *Essential Statistics for Medical Examinations* which offers basic explanations of some of the more relevant topics. You don't need to read the whole book as it is way too comprehensive for what is required in AKT, but if you pick your chapters wisely, in a similar way to the *InnovAiT* article it might just clarify in your mind exactly what is meant by some of the statistical terms and how they can be applied and are relevant in real-life medicine.

REVISION COURSES

As well as online revision courses there are also several face-to-face courses on offer. Those accredited by the RCGP can be found under the 'Events' section on the RCGP website and cost around £165 to RCGP members for a one-day course (£225 for non-members). There are other companies which are also providing preparation courses for AKT, such as Emedica (http://emedica.co.uk) which charge around £195 for a one-day course, although the credibility of these courses is uncertain.

Just a note on the GP Update courses which some of you may have had the opportunity to attend. These are normally based at the level of qualified GPs and are a way of consolidating the evidence published in recognized journals in order to bring GPs up to date on current evidence-based practice without them having to spend hours searching through journals. If you have the opportunity to attend one of these courses prior to your AKT, bear in mind that some of the evidence discussed may not yet have been implemented in the NICE/SIGN guidelines etc and therefore for the purposes of the exam, your answers should be based on the most recent recognized nationally published information. The chances are if there has been a recent significant change in practice that could cause controversy regarding the correct answer in the exam then this question is likely to be omitted.

What books should I buy?

In my opinion, there isn't any one essential book that you should buy in preparation for this exam, although there are plenty of AKT revision books available via the RCGP bookshop. You may want to consider buying the *Oxford Handbook of General Practice* if you haven't done so already, but if you'd rather practise some questions from books then it might be worth buying a selection between a group of you and swapping them once you have completed all of the questions.

Should I have some time off for study leave prior to the exam?

Many people do, as it gives them a chance to complete plenty of practice questions and consolidate the knowledge that they have been brushing up on over the previous few months. It's sometimes helpful to spend some time getting your head around the national guidelines for things during this time, as very often in practice you are adhering to your local policies which may be very different. Unlike CSA, missing out on a few days in surgery to cram for your AKT is not really going to do you any harm, so if you can get the time off then do so, but use it wisely!

On the day

What do I need to take on the day?

The only two things that you need to bring are the all-essential identity documents. One should show your name, photograph, and signature (e.g. passport or driving licence) and the other should show your name and signature (e.g. credit card). Make sure that the name on these documents is the same as the name in which you are registered with the college, because if the names are different you will be refused admission to the examination. This is because Pearson VUE must be certain that the person sitting the test is the same person who is registered with the college. If for some reason your name has changed since registering with the college then you must provide evidence of this change in the form of a an official document such as a marriage certificate.

More information regarding this will be detailed in your confirmation email from Pearson VUE.

Tip: Don't forget your **two** essential identity documents:

- Passport/driving licence
- Credit card

What happens on the day?

When you arrive at the Pearson VUE centre you will be asked to show your identity documents (which must match your details from your application) and will then be ushered in to a holding room until all the candidates have arrived. You must place all of your belongings in a secure locker and cannot take any equipment in to the exam room with you. Refreshments are also not allowed but you can leave water with the invigilator and ask for supervised access to it when required. The room will be monitored by a video link and there will also be invigilators present to prevent any cheating.

What will I find in the room?

Nothing more than a number of individual computer terminals, each with a wipe-clean board for note writing (these must be handed back at the end of the exam).

Can I take the *BNF* in with me?

This is a common question and unfortunately the answer is 'no'. Whilst you may think that this is a strange decision, given that in real life we use the *BNF* as our essential reference book, the theory is that you will not be asked any complex prescribing questions that would require it. Any questions on drugs or drug doses will be related to commonly used medications in their normal doses, which you should know based on your day-to-day experience in general practice.

And no, you won't have access to 'GPNotebook' either—sorry!

One of my colleagues has said that we need to memorize parts of the *BNF*—is this true?

Most definitely not! See previous question.

After the exam

How do we find out the results?

The results can be found on your ePortfolio under the section 'Progress to certification' on the left-hand side. Once you have clicked on this heading a list of the assessed components of the MRCGP will appear, of which AKT is one of them. The results cannot be found on the generic RCGP website. If you've passed, you'll see a very satisfying little green tick next to the AKT heading (which will say 'Pass') with a magnifying glass icon that you can click on to get a breakdown of your results.

When do we get the results?

The results are on average released about four weeks after the exam on your ePortfolio and tend to come out the evening before the official results date.

What happens if I fail?

If you unfortunately fail the exam then don't despair. You will still receive feedback on your performance which you can then use to tailor your revision for the next sitting. For those starting their training after 1 August 2010 a maximum of four attempts at the AKT will be permitted. If, however, you started your training prior to 1 August 2009 then an unlimited number of attempts are allowed provided that you still have a NTN. Think carefully about how you could improve the areas in which your performance was poor and discuss your educational needs and requirements with your trainer. The chances are that you will fly through the next sitting so don't give up hope!

Do we get any feedback on our performance in the exam?

Yes but not in the specific sense, although the RCGP do provide general feedback for each part of the exam to discuss areas which were in general answered badly. When you click on the magnifying glass icon in the AKT section of your ePortfolio, you will be taken to your breakdown page which tells you your overall score compared to the mean mark and the pass mark. The 'mean mark' is the average mark scored by all candidates taking the exam (remember to brush up on mean/median/mode for the exam!) whereas the pass mark is the threshold above which you are deemed to have passed the exam. It will then give you a breakdown as to how well you did in each component of the clinical medicine, evidence interpretation, and organizational questions.

The RCGP reports on the AKT can be found under the MRCGP section of the RCGP website and go as far back as the October 2007 sitting. It's well worth having a read through these before the exam to find out areas which have been answered poorly in the past and to find out some of the common mistakes that candidates were making.

How long does an AKT pass last for?

AKT passes obtained after 1 August 2010 are no longer subject to a three-year validity limit which they were prior to this date. This had implications for those who took time out for maternity, paternity, or sickness reasons. Similarly, AKT passes obtained between 1 August 2007 and 31 July 2010 will remain valid pending the award of your CCT.

Additional questions?

If after reading this chapter you still have some unanswered questions about the AKT, then scribble them down so that you don't forget, and ask your trainer or course organizer at your next VTS teaching session. The chances are that someone else will be pondering over the same thing and the only way to find out is to ask!

And finally... **good luck!**

REFERENCES

1 RCGP. *Applied Knowledge Test*. Available at: http://www.rcgp-curriculum.org.uk/nmrcgp/akt.aspx (accessed 12 August 2010).

2 RCGP. *AKT presentation for candidates*. Available at http://www.rcgp-curriculum.org.uk/nmrgcp/akt/presentations.aspx (accessed 12 August 2010).

3 Irving, G (2009). Goodbye to goobledegook: an introduction to basic statistics in primary care. *Innovait*, **6**, 372–83.

4 RCGP. *InnovAiT*. Oxford University Press, Oxford. Also available at: http://rcgp-innovait.oxfordjournals.org/ (accessed 12 August 2010).

CHAPTER 15

Life events and changes that may affect your training

On the day that you sign up for your GP training rotation you will have no idea what life has in store for you over the next few years and as a result there may be times when things change, become very difficult or more stressful, or don't go as planned. That may be due to unfortunate things such falling ill, family bereavement, financial difficulties, or training problems or may be a result of exciting things such as getting married, moving away, having a baby, or winning the lottery. Either way, it would be fair to say that not only could these situations affect your personal life and well-being, but they may well influence the length, success, or enjoyment of your training. This chapter therefore aims to talk through some of these scenarios, in the hope of limiting the potential effects on your training as well as pointing you in the direction of the right sort of help and support should you need it.

The bad stuff

Becoming ill

Even though as medics we know that illness can affect anyone at any time, it is often one of those things which you never think will happen to you, especially during your twenties and thirties. But you don't need to get a serious illness in order for sickness absence to have an impact on your training and in fact it's the minor things such as broken bones and bugs which are more likely to cause the hassle. Whilst these sorts of things normally mean that you are only away from work for a matter of weeks, those weeks plus any other one-off sick days you may have had, can suddenly start to add up leaving you with not enough days of experience to count towards your training. The PMETB order states that whilst the training is now competency based, you are still required to complete the full three years' worth of training with no allowances made for jury service, sickness absence, or maternity/paternity leave[1]. The RCGP will, however, make allowances for a maximum of two weeks absence over each calendar year but any leave taken in excess of this must be made up in full (although not necessarily in the post in which the absence occurred)[1]. You are also required to inform the Director of Postgraduate GP Education within your deanery of any significant absences which you may have had or be intending to have, as the compliance with absence from training is managed at a deanery level.

In terms of sickness pay, you will be entitled to the same amount as equivalent hospital doctors (i.e. it depends on the number of years of NHS service that you have provided plus the number of months of sickness leave that you will have taken[2]). A summary of the entitlements is shown in Table 15.1 but if you require more information on sick pay then this can be found in **CHAPTER 19, 'THE JOB RACE!'**. This chapter contains a large section on salaried GPs and discusses in detail the BMA model contract which has very similar terms to that of the GPST contract.

Table 15.1 Sick pay entitlement for GPSTs

Year	Full pay	Half pay
First (once >4 months service completed)	One	Two
Second	Two	Two
Third	Four	Four
Fourth	Five	Five
Fifth	Five	Five
Sixth	Six	Six

Some of you may also have private income protection policies to supplement your income should you be off sick for any length of time. Whether or not you should consider one of these polices is up to you and as with any type of insurance it offers 'peace of mind' when considering a situation that will probably never occur.

Tip: Remember that if you have more than two weeks' absence in any one calendar year then you will be required to make up that time by way of extending your training.

Not getting on with your trainer or educational supervisor

Hopefully, this shouldn't be a problem which occurs very often but it would be unrealistic to expect all GPSTs to form good relationships with their trainers or supervisors. If you are one of the unlucky few who may be affected by this issue however, then it's important that you don't struggle on alone. Whilst putting up with an awkward supervisor for one of your six-month posts isn't the end of the world, if you have a serious personality clash or similar issue with your trainer or educational supervisor then life could become a little more unpleasant, as you will be required to spend a fair amount of time with that person. A poor relationship between a trainer and a trainee can have devastating consequences for both parties involved and it is therefore essential that you address any issues very early on, in the hope that they can either be rectified or that you can be assigned an alternative trainer or practice. No one can be expected to get on with everybody and whilst in theory you should only need to meet that person on a professional basis, most trainees develop a more supportive relationship than that with their trainer and it would be a shame for a trainee to instead dread every necessary contact that they have. For those who are having issues it can, however, be extremely difficult to speak out when things are going wrong, either for fear of being accused of a 'whinger' or because of concerns regarding what type of reference you might get when it comes around to applying for jobs. The following questions may help you to think about whether the issues you have with your trainer/educational supervisor are rectifiable:

1. Has there been a specific incident which has affected the relationship between you and your trainer/ES?
2. Do you feel that you can speak to your trainer/ES about the problem?
3. Do you generally feel supported by your trainer/ES?
4. Do you feel that you are getting the right type and amount of teaching that you require?
5. Are you otherwise happy within your post?
6. Do you feel intimidated or threatened by your trainer/ES?
7. Have you lost respect for your trainer/ES?
8. Do you feel that you are being taken advantage of within the practice?

If the answer to the first five questions is 'yes', then it is likely that the problem that currently exists is rectifiable once you have had time to sit down with your trainer and discuss the relevant issues. Sometimes it is just a case of getting off on the wrong foot and if both parties are willing to discuss the matter in a professional and non-judgemental way then there is no reason why you cannot go on to re-build the relationship with your trainer/educational supervisor with the hope of a positive outcome. If, however, the answer to the last three questions is 'yes' then it may well be that the relationship has irreparably broken down between you and your trainer and you should therefore discuss with your course organizer the prospect of moving to an alternative practice for the sake of both your training and general well-being.

Have the confidence to speak out. Whilst you are there to do a job and fulfil a role, it is not just about service commitment and you are entitled to receive the required amount of support and teaching within your training post.

> 'The first few months of my ST3 placement were really tough, as I found that there was a real clash between how I liked to learn and how my trainer liked to teach. He was totally unaware that his method of 'making me stand on my own two feet' from an early stage, was actually having a detrimental effect on my confidence and general happiness. It wasn't until we sat down and talked about how I was feeling that he realized how perhaps a different teaching style would be more effective. Given my initial unhappiness I was given the option to move practices but after we'd had the conversation there was really no need as things improved dramatically after that point.'
>
> ST3

Having second thoughts

If at some point during your GP training you begin to worry whether or not you have made the right decision in choosing general practice then please don't panic—you certainly won't be the first trainee to feel like this and you definitely won't be the last. Often it can be a result of not enjoying your current placement or perhaps due to a change in personal circumstances, but the chances are that it is likely to just be a blip, a bit like a mid-life crisis where you suddenly begin to wonder whether the grass is greener on the other side of the fence. The training years are hard work and you may often find yourself wishing to be rid of the stress of the ePortfolio and the anxiety related to exams. Having been there and done that I can assure you that it is all worth it in the end; however, that's not to say that you should make your life miserable just so that you can get the letters MRCGP after your name.

Now that most junior doctors enter the training programme at the end of the FY2 year, there is very little time for you to really think about whether this is the right career for you and whether it is something that you want to do forever. Well, the first thing to say is that it doesn't have to be forever and the second thing to say is that you can do almost anything with a general practice qualification. And if you decide halfway through your training that you don't even want to go as far as completing your membership then you aren't a failure. Many people have second thoughts at some point during their training for one reason or another, and end up living much happier lives in either an alternative field of medicine or by doing something completely different. Although there are likely to be more opportunities for a doctor with membership to a recognized college, there are still plenty of jobs out there for fully registered doctors who wish to consider a slight career change. Examples of these include working for the Ministry of Defence, cruise ship medicine, prison medicine, work with the pharmaceutical industry, and providing medical advice for the medical indemnity organizations. The Internet provides a wealth of information on the different types of opportunities available, although if you really are having serious second thoughts about completing your training then you should ideally discuss this with your educational supervisor or one of your course organizers. Sometimes the option of swapping a particular post within your rotation or moving to an alternative practice may be the little change that you need to make you happier in your training.

> Tip: Sometimes the option of swapping a particular post within your rotation or moving to an alternative practice may be the little change that you need to make you happier in your training.

If you're just having a wobble and need to have some reassurance about your career choice then try looking back at **CHAPTER 1** and remind yourself of the reasons why you chose general practice—it might just be enough to convince you to keep going. However, if you really are struggling with second thoughts about general practice then please you don't suffer on in silence—tell someone early on and it might just be that things can be changed to make you feel better.

Failing the exams

Whilst this is an issue which no one hopes that they will have to contend with, unfortunately the very nature of the beast is that some people will fail their exams at some point and others may not ever manage to pass the required elements of the training scheme. If you have made the best possible efforts in terms of attendance, WPBA, and completing your out of hours then you are likely to be looked on more favourably at panel than if you have consistently done the minimum to get by throughout the scheme. This doesn't mean though that you will be 'let off' having to pass your exams as these are a professional requirement. It might, however, mean that you are more likely to be granted an extension to your training to allow you to resit the necessary parts of the assessment process. It isn't a given that this will happen and therefore you need to try and make the most of the time that you are guaranteed a training job.

Should you be one of the unlucky ones that fails any aspect of the scheme then don't panic. There are plenty of people who will be willing to help you get through as long as you are willing to make the effort and take on the responsibility of getting the required elements done. Some schemes will also offer specific mentors for failing trainees who can provide them with the necessary guidance and one-to-one support that they need to get through.

The good stuff

Starting a family

Having a baby is undoubtedly one of the most exhausting but exciting and exhilarating experiences that you can have, so it would be unrealistic to think that it wouldn't have an impact on your training. Lugging around a home visit bag when you're 37 weeks' pregnant or trying to sit CSA when you've been up all with a poorly baby the night before will obviously be very difficult, but it is not just the practical aspects which can be problematic. Your training programme is likely to have to change with alterations made not only to the posts that you had planned to do but also to the timing and number of required WPBA assessments. Your CCT date will inevitably be delayed and you will also have to cope with the additional stress of supporting your family whilst receiving maternity pay (although for GP trainees you actually get a pretty good deal on this one!).

However, it doesn't have to be a stressful and difficult experience (and in fact many TPDs will encourage you to start your family during your GP placements!). As long as you are well prepared for the changes that will happen and organize yourself so that you are still able to complete the necessary components of the training programme, then there is no reason why your training cannot continue to run as smoothly as before. For more information on maternity leave, maternity pay, and going back to work as a LTFTT (less than full-time trainee) see **CHAPTER 16, 'HAVING A BABY AND BECOMING A LTFTT'**, which hopes to provide you with all of the essential information that you will need should you find yourself pregnant during your GPST rotation.

Moving house

Whilst in an ideal world you should complete the three years of your GP training within the same deanery, there will undoubtedly be circumstances for some in which they are required to up and move to a new location and hence an alternative deanery. Transfers to a different deanery, however, are at the discretion of the accepting Postgraduate Dean and there is no automatic entitlement or right for this to occur. It is normally only considered in exceptional circumstances and when you have been in the programme for at least one year. Trainees who need to move for ill-health reasons or those who have responsibilities as a carer will get priority. Of course, it also depends on whether there is suitable post or programme available within the receiving deanery.

The move will need to be coordinated by the home deanery and if, for example, it is required as a result of marriage, birth of a child, or illness you will be required to provide evidence of this. In order to apply for a transfer you will need to complete the relevant IDT1 form from your deanery and send it along with an up-to-date CV, written details as to why the move is necessary, confirmation of support from your TPD, a complete assessment record (including the most recent ARCP panel outcome), and two references submitted using the structured reference form.

Applications will be considered during two 'windows' within the year, namely May and November when the transfer panels take place. If you are accepted by the receiving deanery but there are no vacant posts then you will need to apply again during the next 'window'. If for some reason your application is declined then you do have the right to appeal but there is no guarantee that you will be allocated a post.

Moving to work in a different location can bring many new challenges in terms of getting to grips with local referral systems, finding your way around, liaising with unknown consultants, losing your current support network, and even understanding the local accent!

'Moving house is difficult. Be prepared for everything to take longer than before, particularly home visits! Becoming familiar with your practice area takes time. A sat nav is an essential, but also get an A–Z, as it will mean that you become less reliant on the sat nav and start to get a clearer picture of the overall area. I have also found that the interface with secondary care has been more difficult than I imagined. It was always nice as a GPR to know who you were referring to, and what the process is on the other end. It also makes private referrals difficult if the patient hasn't a clear idea of who they want to see. I am starting to try and go to educational meetings with hospital consultants to try and build up some local connections and be able to match a face with a name.'

Newly qualified GP who moved to Scotland to take up a partnership

Taking some time out

If you're feeling that the long road ahead is all seeming a little too much right now and that dream that you had for travelling the world and broadening your horizons seems ever so much more unlikely than it did last year, then perhaps you should consider doing an Out Of Programme Experience (OOPE). Seeking to take time out of the training programme may be requested for a number of different reasons including travel (OOPE), research (OOPR), a career break (OOPC), or clinical experience elsewhere that has been approved by PMETB for training (OOPT); however, in each case it must be agreed by the Postgraduate Dean for the area. It is not normally granted until you have been in your training post for at least one year and normally depends upon there being exceptional circumstances.

Should you decide that this is something that you would like to do then you would need to complete the Out Of Programme (OOP) application form available from your deanery. The process can take a considerable time to complete and be approved, therefore it's important that you apply as soon as possible in order to get the ball rolling. Not all applications will be granted and there normally has to be an exceptionally legitimate reason for your request. However, don't be put off by the lengthy approval process—if it's something that you feel strongly about then commit to your cause and you hopefully should get the time out that you need. A maximum of three years can be taken out of the training programme and you will need to give six months' notice of the date which you plan to return to the scheme. The placement that you are given will depend upon the availability at the time and it may well be that you will have to wait for a suitable placement to appear. For each year that you are out of the programme you will need to return an annual OOP report which will be considered by the ARCP panel for review. You will also need to show evidence of the progress that you have made during that year. Failure to do this may result in the loss of your training number.

> 'Having harboured long-held ambitions to spend some time volunteering in a medical capacity in the developing world, my wife and I were keen to take some time out of our training to do this. Initially we knew not whether it would be possible, nor indeed when might be an appropriate time to go without jeopardizing our medical careers. After a lot of research, asking seniors and career advisors, the huge differences in advice and support led us to conclude that such an experience is not well catered for within the current system, and as for when to go...no one had any idea!
>
> Determined to make our ideas a reality, we set about planning our trip, knowing that we would need some support and reassurances to make it possible, both financially and in terms of our careers. We resolved to undertake our 'year out' at the end of my ST1 year which would also be the conclusion of my wife's Foundation training. Once on the GPST programme, I discussed our plans with my educational supervisors and fortunately they were extremely supportive. They signposted me to the 'Gold guide to specialty training', a hefty document which contains a section on applying for time out of the ST programme. Various options are available including applying for time to undertake alternative medical work which can be counted towards medical training, or, as in our case, time on an 'Out of Programme Experience (OOPE)', which would not count towards our CCT (we never intended for our volunteering to officially 'count' towards this).
>
> Six weeks into the ST programme, after contacting the deanery directly and undertaking the appropriate form-filling, my application was initially refused, with the reason being that 'no application would be considered within a trainee's first six months on the scheme'. Positively, however, the Deanery suggested applying later in the academic year. Four months later, I applied again, detailing exactly our reasons for wanting to go, reiterating my subsequent commitment to the ST programme, and explaining in depth about the projects where we intended to spend our time, and what we would be doing. A letter of support from one of my educational supervisors was gratefully attached, and I was afforded a meeting with the Deanery lead, which, after sweating for a couple of weeks beforehand (!), essentially consisted of an informal interview regarding the application. All went well and the application was approved—we were going!!
>
> Since leaving in August, we have had an invaluable experience working at a clinic for homeless children in New Delhi, and are heading on to Nepal and Belize to teach BLS/first aid and a basic medical curriculum to children and staff in orphanages and work at a rural medical

clinic respectively. We will return to the UK early next year in order that my wife can apply for ST training. One key aspect of the 'permission' to undertake our OOPE is that it is important to contact the Deanery in writing half-way through the year to 'reconfirm' my intention of returning to commence my ST2 year in August. Staying in regular contact with my supervisors whilst away has also helped and means I don't feel too 'out of the loop' with current issues.

It seems that the procedure and facility to undertake an OOPE varies widely between deaneries and prior to our application, we had read in the medical press of a couple of failed applicants (who were doing their best to appeal). This was understandably worrying; however, we would attribute our success to a well thought-out and planned application, having ensured that all key parties (supervisors/deanery/organizations) were well-informed in good time, and with good communication throughout.

If spending some time applying your medical knowledge and abilities in a different context is important to you, we would strongly encourage it—it has certainly been worth the effort so far...'

Dr Pete Reeves, ST1

Any changes to your rotation must be agreed by your course organizers and you should fill in an 'unscheduled changes' form and submit it to the deanery. This will enable them to keep your ePortfolio up to date as well as the main trainee database. It also allows them to keep track of vacant posts that may be used for other trainees.

Examples of unscheduled changes are:

- Sick leave
- Maternity leave
- Approved time out of programme
- Changes to rotation

REFERENCES

1 RCGP. *Frequently Asked Questions*. Available at: http://www.rcgp.org.uk/gp_training/certification/faq.aspx (accessed 14 August 2010).

2 NHS Employers. *GP Registrars. Schedules to Direction to Strategic Health Authorities concerning GP Registrars (2003) with 2009 Amendments*. Available at: http://www.nhsemployers.org/PAYANDCONTRACTS/JUNIORDOCTORSDENTISTSGPREG/Pages/DoctorsInTraining-GPRegistrars2007.aspx (accessed 14 August 2010).

CHAPTER 16

Having a baby and becoming a LTFTT

(CO-WRITTEN BY DR SARAH FORBES, ST3 LTFTT, LEEDS VTS)

If you're reading this chapter then the chances are that you are either already pregnant or are considering planning a pregnancy during your training years. If the former of these scenarios applies to you, then firstly congratulations and secondly don't panic. You are certainly not the first person to try and juggle speciality training with the impending arrival of a little one and you most certainly won't be the last. Similarly for those amongst you with a Y chromosome, having a new addition to the family whilst you are still in the process of pursuing your CCT can be an additional challenge and one which unsurprisingly can impact on your training.

This chapter aims to provide a simple yet informative guide for those who fall pregnant during their training years. However, it is by no means comprehensive given that regional variations exist with regard to essential contact details and deanery requirements. More specific local information can be obtained from your deanery but if you'd like to gain more general information on surviving your training during and after a pregnancy, then grab yourself a copy of the book entitled *So you want to be a medical mum?*[1] which manages to cover almost every possible question that you may have.

When is the best time to have a baby?

Ah. The biggest question of all and one which doesn't really have an answer. There is never really a 'right time' to have a baby; however, the best time to consider starting (or making an addition to) your family is when it feels right for you. In an ideal world your occupation and training commitments would not have an influence on this timing; however, in real life this is rarely true. Trainees with experience in this field frequently report that the 'easiest' stage in which to have a baby is during your time spent in general practice, either as an ST2 or more commonly during your final ST3 year. The reasons for this are threefold. Firstly, it is much less hassle trying to arrange your maternity pay through the practice manager of your surgery rather than through the payroll department of a large hospital. Secondly, whilst you are posted in general practice you are unlikely to be required to work compulsory unsociable hours such as evenings, weekends, and nights, which comes as a welcome break given that pregnancy can be tiring enough! You will, of course, still be required to complete the OOH elements of the assessment process but you will have more control over the days, times, and types of work that you do during this time. For example, if possible during your OOH sessions you are probably better off doing telephone triage and base-centre consultations rather than home visits or zooming around in the back of an ambulance with the paramedics. Thirdly, as you enter the third trimester, being able to spend the majority of your day sitting down is likely to be considerably more comfortable than traipsing around on a ward round every day!

> 'In my experience, and speaking to other registrars, it can often feel like you are the first GP Registrar ever to be pregnant. My advice would be to get all the relevant paperwork regarding your maternity leave done as early as possible.'
>
> *ST2*

How much maternity pay will I get?

Maternity pay is made up of two components:

- Occupational Maternity Pay (OMP), and
- Statutory Maternity Pay (SMP) or Maternity Allowance (MA)

Trainees who have completed at least 12 months of continuous employment within the NHS by the 11th week before the expected date of childbirth (e.g. week 29) are entitled to the following breakdown of maternity pay:

- 8 weeks' full pay (OMP) **less** any SMP or MA and then
- 18 weeks' half pay (OMP) **plus** SMP or MA and then
- 13 weeks' SMP or MA

Those trainees who haven't completed 12 months of continuous service with the NHS by this time will therefore not be entitled to NHS OMP but may still be able to claim SMP or MA, as these payments reflect the national insurance contributions that you will have made over the years and are not paid out of your employer's pocket.

Maternity pay is calculated on the basis of your average weekly earnings for the eight weeks leading up to the qualifying week, which is the 15th week before the expected date of delivery (e.g. week 25). Gross earnings are taken into consideration (including banding supplements) and therefore it is important to ensure that you are earning your **normal** salary during the maternity pay calculation period, as this will determine the amount of maternity pay that you will receive.

Who will pay my occupational maternity pay?

OMP is usually paid by the employer for whom you are working when you are 25 weeks pregnant; however, the following information should hopefully make it clear with regard to who should be paying you what.

Working for a primary care trust (PCT)

If you are working in a GP surgery at the time of being 25 weeks pregnant and will still be working for them when you go on maternity leave, then it will be the PCT who are responsible for paying your OMP.

Working for a hospital trust

If you are working in a hospital post at the time of being 25 weeks pregnant and will still be working for them when you go on maternity leave, then it should be the hospital trust that pays your OMP.

Problems, however, can occur if you are unfortunate enough to change posts during the course of your pregnancy. In the event that this arises then you should get your OMP from whoever you are working for at the time of going on leave.

Regardless of who is paying your OMP you will need to provide them with a signed copy of the MATB1 form given to you by your midwife. Make sure that you not only keep a copy of this but that you also keep the original in a safe place. The MATB1 is also required if you are applying for MA rather than OMP and SMP.

> Tip: Keep a folder with all your maternity paperwork in, including your wage slips from the beginning of your pregnancy. This will make it much easier when filling in all the relevant forms.

What's the difference between Statutory Maternity Pay (SMP) and Maternity Allowance (MA)

Essentially in monetary terms there isn't really any difference between these two payments; however, this can cause considerable confusion when applying for your maternity pay. If you have been employed by the **same employer** for 26 weeks by the 15th week before the expected date of delivery (e.g. week 25) then you are entitled to SMP. This is automatically added to your pay slip and comes directly from your employer. Essentially what this means is that at your EDD you will still be employed by the same employer that you were employed by at the time of falling pregnant (e.g. you have worked in that post for at least 6 months). This situation most commonly occurs during your ST3 year where you are fixed to one post for at least a year; however, if you are working at 'less than full time' it may be that this is also the case in some of your hospital or ST2 GP posts. For SMP purposes therefore, the rules about working for the 'same employer' actually mean that you need to have been paid by the same person for the last 26 weeks rather than just that you have been working for the NHS. This makes things a little more complicated for those who may change posts in the middle of their pregnancy, however they should not lose out financially. Whilst you will not be eligible for SMP you will instead be able to apply for MA which is paid directly via the government. The monetary value of this is the same as SMP, however it relies on you gaining an SMP1 form from your employer. This form gives an explanation as to why you are not entitled to SMP and forms a part of your MA application. Currently both SMP and MA are paid at £124.88 per week (as of May 2010).

> Tip: Remember that during your training years, whilst you may have completed at least 26 weeks' continuous service with the NHS, you may not have completed 26 weeks with the same employer by week 25 of your pregnancy and will therefore need to work out whether it is SMP or MA that you are entitled to.

How do I apply for maternity allowance?

This is done by completing the MA1 form which is available from Jobcentre Plus, or can be downloaded via the DWP website[2]. You will also need to submit three pay slips from the preceding year plus the SMP1 form which states why you are not entitled to SMP.

What happens in terms of maternity pay if my due date is close to my CCT date?

If you reach week 25 whilst still in your ST3 post and are planning to qualify at the usual time then you will still be entitled to receive SMP but will not get any OP. You will, however, have to start your maternity leave earlier than you might have liked to in order to qualify for the money. Interestingly though, you are entitled to work as a self-employed doctor whilst receiving SMP without it affecting the payments that you receive.

In order to qualify for OP as well as SMP, you must have reached week 29 before the end of your training.

Who do I need to inform about my maternity leave?

In the early stages of your pregnancy it may be wise to let your clinical supervisor or GP trainer know so that there is someone on site who knows that you are pregnant for health and safety reasons.

In the later stages, however, you are obliged to provide your training programme director and your payroll department with the following information in writing by the end of the 15th week before the expected date of childbirth (e.g. week 25):

1. Your intention to take maternity leave
2. The date you wish to commence maternity leave (you may not know the exact date initially but ideally you should give your employer a vague idea of your leave date in order for them to organize cover during your absence)
3. Your intention to return to work after your maternity leave (you need to return to work for a minimum of three months to be entitled to full occupational maternity pay)
4. A copy of your MatB1 form to confirm the above (this is available from your midwife)

You should also inform the deanery so that adjustments can be made to your ePortfolio and your panel reviews can be adjusted accordingly. Some deaneries will ask that you complete an 'out of programme' form which will need to be signed by one of your TPDs to confirm that you will be taking time out from the scheme.

> Tip: The key to success is to be organized and start looking into financial matters early! This will hopefully lead to a stress free maternity period and avoid endless phone calls getting passed from pillar to post whilst trying to pacify a crying baby!

What other benefits may I be entitled to?

1. **Health in Pregnancy Grant**
   This is a one-off tax free payment of £190 for anyone who is at least 25 weeks pregnant. This is unfortunately being phased out from January 2011.

2 **Child benefit**

Currently all parents are entitled to receive child benefit regardless of your salary. You can apply for this as soon as you have the birth certificate; however, it is worth printing off the forms[3] and getting them ready beforehand as the first few weeks with a baby can be pretty chaotic!

3 **Childcare vouchers**

Childcare vouchers enable employees to sacrifice some of their salary to go towards nursery fees. The maximum value that be used in this way is £243 per person per month; however, it is possible for both parents to commit to this scheme if their employers allow. By doing this you gain a small amount of tax relief, as the selected amount is deducted from your salary prior to calculating your required tax payment. Whilst this scheme still exists at present, there was talk in 2009 that this would change, so bear this in mind when planning your finances. Not all employers sign up to the childcare vouchers scheme, however big organizations such as the NHS and the PCT should do. When you move out into general practice as a fully qualified GP, however, it may be that you are not able to benefit from this scheme.

If you are planning to utilize the childcare vouchers scheme then make sure that you set it up as soon as you possibly can as it is often a lengthy process. There are many different companies which provide the nursery vouchers service so contact your employer in the early stages to find out exactly what is required of you. Most companies will want to know the details of the nursery or childminder that you are employing as they will not pay out to anyone who is not a registered child carer (e.g. they won't pay your mum for doing the job unless she's properly qualified and registered!).

4 **Reduced subscription fees**

Prior to commencing your maternity leave it is worth phoning the companies with whom you hold subscriptions to see if you are entitled to reduced rates whilst you are not working. You should also let the RCGP AiT office know as they may suspend your ePortfolio and alter your AiT fees.

How much maternity leave can I take?

You are entitled to take up to 12 months of maternity leave from the date that your child is born (regardless of your due date), however it's worth noting that as you still accrue annual leave during this time you may therefore may be able to take more like 13 months out of work.

Can I take time off for antenatal appointments?

Yes, you are entitled to paid time off for antenatal care, however it is essential that you give your employer as much notice as possible. If possible try to organize your appointments at a time when it is less busy on the ward or at your practice; however, there will obviously be circumstances in which you have no choice in the matter given that your midwife is likely to only be available at certain times.

Will it affect my rotations?

Yes—it might do. If you are due to reach week 29 in your current job then you will not rotate to your next placement at the usual time. Instead your employer is obliged to extend your contract for maternity leave purposes. You will then rotate to the new post on your return to work. This prevents you from leaving a post having only just started and makes working out who should pay your maternity pay much easier!

How long will I have to do on calls for?

This will be different for each individual. If you are in a general practice placement then you may feel that you are no longer able to do your OOH commitment once you become heavily pregnant. It is best to discuss any issues such as this with your trainer. Your pay will not be affected if you work a reduced number of hours due to pregnancy-related problems, however your employer may ask for a note of proof from your own GP.

On calls are more of a problem for women in hospital posts and the basic answer is just to 'see how it goes'. If you're feeling really well and are prepared to continue with your on-call commitments as normal then that is fine, but there is no point trying to be a martyr and 'solider on' if you're finding things tough. No one will thank you for struggling on and it's important that you put the health of you and your baby first. It is best to try and get a balance between playing it by ear and giving the rota coordinator enough time to cover your shifts. If there's any doubt though then opt out and let them arrange a locum as this is much more preferable than ringing up just before your shift to say you can't do it. Again, you may be required to bring a note from your own GP but most GPs are quite happy to do this. You will still get paid your banding even if you are not doing the on-call component.

Ideally you should undergo a risk assessment with the occupational health department once you have announced your pregnancy to determine whether there are aspects of your work that are no longer appropriate. This, however, will depend on the job that you are doing and the area within which you work and is likely to vary hugely across the country.

What if I become ill during my pregnancy?

If you are off sick before the 36th week then you are entitled to sick pay as normal. If, however, you are unable to work due to a pregnancy-related problem after week 36 then you will have to commence your maternity leave and the clock starts for your maternity pay and leave allowance.

I've heard that you still accrue annual leave whilst on maternity leave—is this true?

Whilst it may sound like a strange phenomenon, yes, you are still able to accrue annual leave whilst you are on maternity leave. It is also worth being aware that it will be accrued in a proportional way to the amount of hours you were working before you started your maternity leave. For example, even if you only plan to go back to work part time but had been working full time up to the date that you went on maternity leave, then you will accrue annual leave at full-time proportions. Your annual leave, however, cannot be carried over to the next training year and is paid prior to your return to work. This may mean that for the last month of your maternity leave you get paid in full.

What about paternity leave?

All expectant dads are entitled to two weeks of paternity leave although it's worth noting that if during that same year they have had any other absences due to sickness etc. then it may mean that they will go over their maximum limit of two weeks out of training per year. This may result in a small extension to their training in order to comply with PMETB.

What do I do about my ePortfolio?

Prior to commencing your maternity leave you will need to inform the ePortfolio administrator within your deanery. This will enable them to document the information within the 'posts' section of your ePortfolio so that anyone accessing it can see that you are legitimately on leave and therefore won't give you hassle if there are no log entries submitted during that time! Maternity leave should then appear in your 'posts' section and the ARCP clock stops. Make sure that you have an up-to-date review within the last six months and have completed the necessary WPBAs prior to going on maternity leave in order to prove that you are making progress with the ePortfolio.

What are 'keep in touch' (KIT) days?

KIT days were developed in 2007 so are essentially still in their infancy. They were designed to enable employees to keep in touch with their workplace and ease themselves back into their previous role without it affecting their maternity pay or leave allowance. You are permitted to complete up to 10 KIT days without it affecting your maternity pay and they do not have to be consecutive. They also don't necessarily have to be days in which you do clinical work and can instead be study days or teaching. You should get paid at a daily rate which can be negotiated with either your practice manager or payroll department, however it's probably better to agree terms beforehand rather than having to sort it out after you've done the work.

What if I want to work part time when I return to work after maternity leave?

Going back to work part time is known as working less than full time (LTFT).

Becoming a LTFT trainee (LTFTT) involves working a reduced number of hours compared to your full-time colleagues. This can be for a variety of reasons, however the most common reason for this is people returning from maternity leave and needing to provide child care for some part of the week. Most LTFTTs will be offered a contract at 50% of full time, to enable the deanery to organize a job share arrangement; however, it may be that local variations in this will exist. If you are planning to return to work part time after maternity leave then you should ideally give the deanery at least 12 weeks' notice in order to organize this.

How many hours will I work as a LTFTT?

The number of hours will depend on the contract you have agreed to. This will be negotiated with your deanery and TPDs but in the majority will equate to 50% of full time. Your timetable should look similar to that of your full-time colleagues (see **Chapter 11, 'Your first GPST post and some basic tips on how to survive'**) but obviously at a proportionally reduced amount depending on what percentage commitment you have agreed to in your contract.

One way of looking at it is that you are doing a full time job but over two weeks. This is useful when designing your timetable as you can see what your full-time colleagues are doing and then work out the same amount of sessions. This ensures that you get the same amount of clinical and tutorial/teaching time even though you are working less than full time.

Do my ePortfolio minimum requirements change when I convert to becoming a LTFT?

Unfortunately not. In the main you are still required to complete the same bulk of evidence and learning log entries as your full time colleagues. This therefore requires you to be extra organized in order to fit them in within a reduced number of sessions. Start thinking about your assessments at the beginning of each placement and plan when you are going to achieve them so that you allow yourself enough time with your clinical and educational supervisor.

> Tip: Organization is the key to success. As a LTFTT you need to be more organized than your full-time colleagues as you often have the same amount of work to do in a reduced amount of time... and just because you're not at work doesn't mean you're not working, being a mum is often even busier!

Do I have to do the same number of out of hours sessions as the full-time trainees?

Currently the out of hours requirement is pro rata (e.g. if you are on a 50% contract then you do 50% of the out of hours as full time trainees). There may be slight regional variations in this however, so it is worth double checking with your deanery or TPDs.

> Tip: Keep a note book in your bag and whenever you have any teaching jot down a few notes. As a mum it's amazing how quickly 'baby brain' takes over and the session is forgotten as soon as it's over! Every so often look back through your book and enter the teaching as a log entry to refresh your memory.

When do I need to see my educational supervisor?

You are required to see your educational supervisor every six months, which is the same as full-time trainees.

How much study leave and annual leave will I get?

Your study leave and annual leave will be pro rata to the percentage of work which you are doing. For more information on study leave and annual leave for full-time trainees see **CHAPTER 11, 'YOUR FIRST GPST POST AND SOME BASIC TIPS ON HOW TO SURVIVE'.**

What should I be getting paid?

Given that most LTFTTs will be offered a contract at 50% of full time, then their basic wage should equate to 50% of the pay received by a full-time equivalent GPR. It is best to contact the person who will be paying you (either the practice manager or hospital payroll department) as early as possible in your first post to double check what they think you are contracted to, as this may be very different from what you are actually working.

Make sure that you confirm not only the percentage pro rata basic pay that you will be receiving but also the banding supplement that you should receive. Although the banding may vary depending on the type of post that you are doing, you should still get the same percentage addition to your basic pay as your full-time colleagues. It is much easier to make sure everything is in place in advance rather than getting paid too much and then having to pay it back!

When should I sit my AKT and CSA?

This is a completely individual choice but there are a few things which are worth bearing in mind:

- If you passed your AKT prior to 1 August 2010 then you must complete your CCT within three years of the pass date. If, however, you pass your AKT after this date then you are no longer subject to the validity limits. The same is now true of the CSA
- You can only sit your CSA whilst you are in ST3
- Revision opportunities—trying to revise when you may have little ones demanding your attention is likely to be extremely difficult!

> Tip: If you're planning a family, are pregnant, or have children and wish to join a peer support group with a forum where you can ask questions on all aspects of working as a medical parent then check out the Yahoo group 'Bumps, Babies and Beyond'[4] which was created by a GP Registrar in the Northern Deanery.

What's the big deal about having a family and work–life balance?

Getting your work–life balance right is possibly the most difficult thing that you will be challenged with in life. Being a parent brings its own agenda which you need to try and balance with the commitments that you have made to your training and future career. It may be that you feel guilty when you're at work and not with your children. Or it may be that you feel guilty for actually enjoying being a professional and not thinking about them all day! Getting the balance right can lead to a much happier and healthier life both in terms of physical and mental well-being but it takes time to get your head around it and there are likely to be some stumbling blocks along the way. There are many advantages to being a LTFTT but to fully appreciate these you have to be organized. When you're at work make the most of the teaching and training you receive. Your time with your little ones is so precious that when you are sacrificing this time to go to work you expect high standards of training. The more you put in the more you get out so it's a good incentive to work hard.

In terms of getting work done in your own time, e.g. your ePortfolio or audits, then make the most of friends and family. You may be lucky enough to have family close by. It may be that you arrange a regular slot with another LTFTT where you meet up and take it in turns to look after the children whilst you each catch up with your ePortfolio. Again, it is all about being organized, thinking ahead, and being realistic about what is achievable within the time frame that you have.

> Tip: During your maternity leave try to spend some time researching and visiting your local nurseries and child minders (if you aren't fortunate enough to have family nearby) to help with child care when you go back to work. Knowing that your child is in a safe, clean, happy, and caring environment when you leave them in the morning to go to work makes your day job a lot easier and goes some way to reducing the guilt that you may feel when you first go back to work after maternity leave.

REFERENCES

1 Hill, E (2008). *So you want to be a medical mum?* Oxford University Press, Oxford.

2 Department of Work and Pensions. Available at: http://www.dwp.gov.uk/docs/ma1.pdf (accessed 12 August 2010).

3 HM Revenue and Customs. Available at: http://www.hmrc.gov.uk/childbenefit/online.htm (accessed 12 August 2010).

4 Bumps, Babies and Beyond. Available at: http://groups.yahoo.com/group/bumpsbabiesbeyond/ (accessed 12 August 2010).

PART 4

Early ST3

CHAPTER 17

Out of hours (OOH) and learning to juggle the EWTD again

In 2004, as part of the new GMS contract, GPs were able to transfer the responsibility for providing OOH patient care (that is, outside of the regular surgery hours of 8.30am until 6pm Monday to Friday) to their local Primary Care Organization (PCO). Whilst this came at a cost, it was a welcome change for most GPs who found themselves no longer having to dash out to do home visits late at night, or having their whole family woken by telephone calls in the early hours. Instead, the care was to be provided by specific OOH centres, which would look after a large group of patients collated from a number of local surgeries.

Of course once these changes had taken place it became essential for OOH experience to be incorporated into general practice training, in order to provide trainees with exposure not only to emergency care but also to a different type of patient groups and presentations. Most trainees would agree that whilst there is still a huge variety in what you see during the OOH sessions, it tends to be that acute care forms the majority, as in theory you are seeing patients who are too unwell to wait until the next day to see their own GP. However, you are also likely to see a combination of; patients on holiday, patients who claim to not be able to get an appointment for weeks with their own GP, patients for whom it is just more convenient, patients who want a second opinion, and patients who simply want their symptoms sorted now and aren't prepared to wait until the next day.

Whilst the prospect of having to complete a certain amount of sessions in OOH during your training may seem daunting and an additional stress to your already busy life as a GPST, it is essential in allowing you to gain some experience in managing a variety of patient contacts in a different quantity and context (and it may not be long before the responsibility for taking care of patients OOH is transferred back to the GP surgeries, so it's worth getting a taste of it now!).

'Out of hours' of course isn't to be confused with 'extended hours' which appeared as a DES in 2008, requesting that surgeries extended their opening hours to the evenings or weekends in order to accommodate the working population. The amount of 'extended hours' provided by each surgery was dependent on their practice population and could be scheduled at any time, at the discretion of the practice. You may or may not be required to participate in your practice's 'extended hours' commitment, but if you are then it should be instead of and not in addition to your regular Monday to Friday 'normal hours' sessions.

During your time spent in OOH you are likely to make a significant contribution to the service at what seems to be no cost to them. Remember, however, that there are likely to be financial implications for OOH providers in order to deliver the training and clinical supervision required. Make the most of the sessions that you attend and try not to get too frustrated if it feels like the regular doctors are not pulling their weight. The more patients that you see the more competencies that you will gain and the more confident you will feel at the end of your training.

What counts as OOH?

OOH is classed as 6.30pm until 8.00am during weekdays and all day on weekends and bank holidays.

How many hours do I need to do?

Officially, full-time GPSTs are required to complete a minimum of 36 hours during their ST2 GP post and at least 72 hours' worth of OOH sessions during their ST3 year. This can, however, be spread across the whole of the time spent in general practice equating to around six hours per month. This is relatively easy to achieve if you plan well in advance and get started as soon as you possibly can. Often the weekend sessions are longer meaning that potentially you could get it completed by simply doing a six-hour weekend session once a month. It's also worth appreciating that whilst the college state a minimum hours requirement, it also relies upon you gaining the required OOH competencies during this time. Should your OOH supervisor or trainer feel that you have not had sufficient exposure or experience during your 72 hours' worth of sessions, it may be suggested that you commit to a few more sessions. A six-hour session on a Saturday during which you only see ten patients with coughs and colds (because the football World Cup final is on) is probably not sufficient enough to count as a completed long session. However, if all of the ten patients allowed true testing of your skills of OOH care and allowed you to sign off some of the relevant competencies then that's a different story.

What type of work will I do?

The work that you do will vary from routine face-to-face consultations, to telephone triage, accompanying paramedics, working in a 'walk-in centre', and carrying out home visits. You should try and expose yourself to a variety of different encounters during your time spent in OOH in order to maximize your experience there.

Will I need a formal induction to the OOH facility?

Ideally yes, but it is likely to be brief. Most importantly you should be made aware of any local or in-house protocols, how to get help, where the emergency equipment is kept, and how to use the computer. Most OOH systems will use a computer system which is different from that which you will be used to in practice and it may be worth spending half an hour getting your head around it before you start your sessions.

Will I be left alone?

You should not be left alone in OOH as there should be a qualified trainer available for supervision at all times. There may obviously be times when the trainer is called away to an emergency and you are left alone for a short time but this should be a rarity and not the norm. If you find that you are being left alone on a regular basis then you should raise this as a concern with your educational supervisor.

Do I need to update my professional indemnity?

No. Standard GP registrar indemnity is sufficient for working in OOH.

How do I organize the sessions?

This is likely to vary based on where and when you are planning to do them, therefore the best way of finding out what is happening in your local area is to discuss this with your trainer or TPDs. You will probably be required to submit evidence of your professional qualifications, professional indemnity, recent CRB check, and immunity status before you can be accepted to work there. Approval of these documents can take some time so it is worth planning in advance and finding out what and when you need to send to whom and where (if you know what I mean). The popular sites and sessions are likely to get booked up very quickly as there is likely to be a maximum number of trainees that can be working at one time (not just because of supervision restrictions but also due to lack of available consulting rooms). It is probably best to try and plan your sessions to be evenly distributed throughout the year rather than spending the last few months of your rotation cramming them in and making yourself exhausted.

Will I still be able to comply with the working time directive?

This all depends on the timing and length of the sessions that you are doing but it may be that you are restricted by the EWTD. It is essential that you are properly rested before and after each session worked, whether that is a normal working hours session or an OOH session.

This means that you will need to think carefully about the sessions that you intend to do and negotiate your regular surgery hours with your trainer. You will still, however, be required to complete the same amount of sessions during surgery hours and the OOH sessions are done in addition to and not 'instead of' your normal sessions.

For example, if you chose to do some weekday evening OOH sessions then your trainer may suggest that you to start your surgery later the next morning to allow sufficient rest time between sessions. Essentially, however, it is your responsibility to make sure that you are not trying to fit all of your sessions in within the last few months as not only will this make you non-EWTD compliant but will be exhausting and perhaps be to the detriment of your health and to the patients that you see.

Can I do it in different locations?

Yes—but it is probably easier if you try and do it all within the same PCT. Essentially it will all depend on when and where the available sessions are. Each OOH organization will have their own registration requirements and being accepted for OOH by one PCT does not automatically mean that you can simply turn up and work for another. Recent disasters in the OOH services mean that regulations and demonstration of competence are likely to be tightened up in the near future, so it may become even more difficult to organize the initial registration with each of the organizations.

If possible however, try within your PCT to organize sessions which will expose you to a variety of different learning environments—some centres will be busier than others and offer different services to their patients.

Do I need to be attached to a trainer?

Yes—but this is unlikely to be your regular trainer or educational supervisor. As long as there is a trainer present within the department whilst you are working there then that should be sufficient. The role of the supervising GP trainer is not only to ensure adequate patient care and safety but also to supervise your learning and ensure that you obtain sufficient experience during your sessions. OOH is a good environment in which to get some of your WPBA signed off as if there are quiet periods then the resident trainer may offer to sit in on your consultations and complete a COT or two. You will also probably get the opportunity to spend some time with the nurses in the OOH service so that you can get some experience of dressings and simple wound care which are all relevant to your DOPS.

Will I need to take my own equipment?

This depends on the type of set-up within which you are working but ideally all necessary equipment and accessories should be made available to you. The quality of the equipment on offer however is likely to be variable, so it's probably worth taking your own stethoscope and diagnostic set (otoscope/ophthalmosocpe).

Can I get some of WPBA signed off during OOH?

Yes—and it's an ideal time to get it done. There may be times when there are a few GPSTs working alongside the regular OOH doctors meaning that the resident trainer has plenty of time in which they can carry out COTS, CbDs, and DOPS. You are also likely to experience different types of patients and presentations to your normal routine surgeries so these sessions will no doubt allow you to make valid additions to your logbook and WPBA.

What do I need to learn from the sessions?

The college states that during completion of your OOH sessions you should concentrate on the following competencies:

1. Ability to manage common medical, surgical, and psychiatric emergencies
2. Understanding the organizational aspects of NHS OOH care, nationally and at local level
3. The ability to make appropriate referral to hospitals and other professionals
4. The demonstration of communication and consultation skills required for OOH care
5. Individual personal time and stress management
6. Maintenance of personal security and awareness and management of the security risks to others

Do I have to log all of the sessions?

Ideally for each session that you work in OOH you should complete an associated OOH learning log entry in your ePortfolio. Most importantly this should detail what competencies you have achieved during the session and should also show evidence that you have reflected on the clinical scenarios that you encountered. Try and make sure that each entry is filed against at least one of the curriculum statement headings, the most frequent of which is likely to be number seven on acute care. However, your OOH log shouldn't simply be a list of the hours done and where. You need to show that you are learning from these sessions and show that you have identified further learning needs.

You are also required to get your supervisor to complete a session feedback sheet which you can share with your trainer or educational supervisor. This is available to download via the RCGP website. The feedback sheet requires you to detail the type of session that you attended including the date, time, and length, plus the competencies achieved, any learning needs identified, and any debriefing notes from your supervisor, all confirmed by a signature and a date. This can then be scanned, uploaded, and attached to your ePortfolio as formal evidence of the sessions that you have completed and can be used by your trainer to validate any of the necessary competencies achieved.

What if I don't complete the required number of sessions?

This is most likely to result in a face-to-face deanery panel review and may not be pretty.

Who signs me off at the end?

This is the responsibility of your educational supervisor and not your OOH supervisor. During the final review with your educational supervisor they will be given the option to tick the box which states that you have completed all of the necessary session commitments and have met all of the required competencies. Once this has been done then a green tick will appear in the 'Progress to certification' section of your ePortfolio.

CHAPTER 18

CSA looms… how to stay focused and some tips for success!

Introduction

The CSA is the biggest hurdle which you have to face during the MRCGP assessment process. It's scary, expensive, hard work, and stressful, but is a fair and reliable method of assessing your clinical ability and communication skills. Unlike the AKT, however, you do not have the pleasure of getting yourself comfortable in a driving test centre and having a one to one with a computer. No, instead you have to perform live in front of real-life examiners and real-life patients (well, not quite, but real-life actors pretending to be real patients). It is essentially an assessment of what you should be doing every day in your surgeries at your practice and contrary to popular belief is not a test to try and catch you out. Some of the information in this chapter can be gained by hunting down the relevant documents and links on the RCGP website, but it's sometimes quite useful to have it all in one accessible place that doesn't require an Internet connection or a computer.

General information

The CSA—what is it?

The Clinical Skills Assessment, like the AKT, does exactly what it says on the tin. Clinical skills in the context of CSA are those which enable you to create a rapport and gather information from the patient via both 'talking' and 'doing', in order to assimilate a differential diagnosis and make a shared management plan with the patient. The college, however, has a more official definition which is 'an assessment of a doctor's ability to integrate and apply clinical, professional, communication and practical skills appropriate for general practice'[1].

Where is the exam held?

Unlike the AKT, you do not get a choice of where you will sit the exam, because at present all assessments are held at the RCGP designated assessment centre in Croydon, South London. The exam is held over the 18th, 19th, and 20th floors of a building called Number 1 Croydon, also known as the 'NLA tower', the 'threepenny bit' (if you remember those from 1968), or the 'tower of love' (as we fondly liked to call it) which can be found at 12–16 Addiscombe Road. It sits opposite the entrance to East Croydon Station and on exiting the station you would do well not to notice it.

At present, despite the RCGP headquarters moving to new premises near Euston Station as of August 2012, it is believed that the CSA exam will still be held in Croydon. This may, however, change in the future when the new RCGP premises are complete.

How much does it cost?

At the time of this book going to press, the price for sitting the CSA was around £1,400. This is obviously likely to increase as the years go by as they make slight alterations to the format, but it's a pretty hefty sum of money and therefore might require some saving for!

Who writes the cases?

All of the cases are written by GPs who are currently practising within the UK and who have had specific training by the RCGP in doing so. The cases are mapped to the curriculum statements but are based around real-life situations that you may encounter in British general practice. The part of the patient is played by a role player who is well rehearsed in performing their case and along with the assessor has been given clear instructions as to how the case should run. In fact, each role player is very carefully calibrated with colleagues and examiners before the exam to ensure consistency.

Every case is marked on three domains:

1. Data gathering, technical, and assessment skills
2. Clinical management skills
3. Interpersonal skills

What skills are being tested in the CSA?

The CSA mainly tests from the areas of the curriculum listed in Table 18.1. These areas are then incorporated into the three domains that are being assessed.

Table 18.1 Curriculum areas being tested and their meanings

Curriculum statement (what the book says)	What this actually means
Primary care management	Being able to recognize and manage common medical presentations in primary care
Problem solving skills .	Being able to gather and interpret data so that you can make appropriate management decisions
Comprehensive approach	Showing that you are able to think 'outside of the box' and consider comorbidity as well as health promotion
Person-centred care	How you communicate with the patient and appreciate their health beliefs
Attitudinal aspects	Your ability to practise ethically and appreciate issues of diversity and equality
Clinical practical skills	Being able to competently examine a patient

Applying to sit the exam

How do I apply for the exam?

Application is made via the RCGP website under the section entitled 'MRCGP Assessments'. The college opens to online applications about a month or two before the first week of each exam session, so make sure that you are aware of when the application process opens and closes for your specific exam, as there will only be a short window of time in which you can apply. For example, if you are thinking of sitting the November/December 2010 exam then you will need to have applied via the college website between 28 September and 11 October 2010, with the first week of exams starting on 22 November. Don't try and apply for the exam outside of the allocated time periods, as this is not possible and will only cause stress and frustration.

Once you have booked your assessment online you will receive an email from the college confirming the date and timing of your exam. It will also give you more detailed information on what you need to do on the day. Make sure that you add exams@rcgp.org.uk to your email address book, otherwise further emails may end up being directed to your junk mail folder.

Tip: Make sure that you add exams@rcgp.org.uk to your email address book otherwise further emails may end up being directed to your junk mail folder.

When CAN I sit the exam?

At present the exam sessions run four times a year:

- September
- November
- February
- May

When SHOULD I sit the exam?

This all depends on how much general practice you have done within your training and how confident you feel about being able to consult in ten minutes. The RCGP recommends that you should have completed at least six months of UK general practice before attempting the exam which can only be sat during the ST3 year of your specialist training.

My advice would be not to rush in to sitting this exam. Whilst it may seem tempting to try and get it over and done with early on in your ST3 year, it is an expensive and stressful exam that you should hope to only have to sit once. Whilst I agree that sitting it early gives you the opportunity to re-sit should things not go well, you would be much better to give it a few months and approach the exam with the attitude that you have worked hard, have done enough preparation, and should pass it first time. As with the AKT, you will only 'get it out of the way' if you pass first time and it's an expensive risk to take if you're not sure that you are ready.

> Tip: Don't rush in to sitting the CSA. Make sure that you are fully prepared to pass. You will only 'get it out of the way' if you pass first time and it's an expensive risk to take if you're not sure that you are ready.

Will there be a choice of dates?

Yes, but this depends on how early or late you decide to book the exam. My advice would be to book as soon as the online application process opens, so that you have as much choice in terms of dates as possible. They run from Monday to Saturday, so when booking your exam take into consideration how you are going to get there, how long it will take you, and what other arrangements you will need to make.

> Tip: Book as soon as the online application process opens, so that you have as much choice in terms of dates as possible.

Should I do a morning or afternoon session?

This comes down to personal preference and depends on whether you are a morning or afternoon person. It also depends on how far you have to travel, whether or not you can afford to stay in a hotel, and to some extent what is available on the day that you decide to apply. Leaving it until the last minute may mean that you are left with the Friday afternoon or Saturday morning sessions! For some it may also depend on other factors such as childcare arrangements or availability of transport, so it's worth thinking about well in advance.

As with AKT, those who sit the exam in the morning will be kept in quarantine until the afternoon candidates have been safely locked away in the holding room, so you are unable to escape immediately after the exam is over.

What is the best day of the week to sit the exam?

Who knows?! Unless you have funny ideas about Friday the 13th or other strange superstitions, it really doesn't matter. It's amazing how many people chose to do the exam on a Saturday, when at least if you do it during the week you get a free day off work! You could argue a case for every day of the week and essentially everyone is different, but if you ask people why they chose a specific day then they will normally have some strange, complicated (if not slightly unethical) or amusing reason for why they did so.

> 'I decided to do the exam on a Wednesday morning as it gave me Monday morning to get back in consultation mode after the weekend, Tuesday off work to pack bags and navigate my way down to Croydon on the train, Wednesday morning for the exam followed by lunch and a large glass of wine and because Thursday was our VTS half day it meant only half a day at work before getting the chance to have a good old debrief with the rest of the gang in the afternoon about how it went!'
>
> ST3

Preparation

What types of cases will there be?

This is a commonly asked question and candidates will often probe trainees in the years above to tell them what specific cases they encountered in the exam and what things they should focus on during revision. Essentially what you need to remember is that no one is trying to trick you and that the types of cases that you see will be very similar to those which you are encountering on a daily basis in your surgeries. In fact, it's even better than that, because they should really only have one problem and will definitely leave after 10 minutes (which I bet you wish was the case with some of your heart sink patients!).

All of the cases are played by professional actors who are role playing typical presentations that you may encounter in UK general practice. They will respond to your consultation style in the same way that a regular patient would, but will not reveal their whole script unless you ask the relevant questions. They are advised on how to respond to certain questions and behaviours exhibited by the candidate and have been well calibrated in these responses prior to the exam. They are not there to try and trick you but at the same time cannot give you additional clues, nods, or winks like they might have done during your medical school finals.

Patients who appear to have lots of problems on the background information sheet will still have only one problem that needs addressing during that consultation and it may be that the case centres on how you deal with medical complexity and comorbidity. Remember, unlike real life, you don't have to worry about getting your QOF points, although you might want to consider opportunistic health promotion if it is relevant and you have the time.

If you do, however, want to have a basis of topics for your revision then it's probably worth looking at each curriculum statement and pinpointing the common and important conditions that are listed. The CSA circuit aims to involve a mix of cases in order for it to be a fair assessment and therefore the street-wise candidate will think about what is common, testable, and will add variety. Think about some of the more challenging cases that you could encounter, e.g. more than one patient, reluctant patients, breaking bad news, angry patients, and those that turn up with a list! You will undoubtedly also get a case based around a medical emergency and whilst you won't be expected to follow ALS guidelines and pretend to resuscitate someone that has just re-enacted a fantastic cardiac arrest in your consulting room (although if they do, it's probably for real and you should act accordingly!), you should, however, be comfortable with recognizing a medical emergency and allow yourself to be appropriately doctor-centred in your management!

> Tip: Remember—CSA isn't just a test to see whether or not you can be nice to patients like the GP exams were at medical school. You are real doctors now after all and you need to make it clear that you know what you are doing clinically as well as having good communication skills!

Finding out which cases have been used in past exams from fellow trainees is a massive DON'T! The RCGP now have thousands of questions in the case bank and the chances of you getting the same case as your mate that did it last week are slim to none. Sharing case information is taken very seriously by the college and prior to sitting the exam you will be asked to sign a declaration stating that you will not share any of the information regarding the exam. Failure to comply with this may lead to irreparable consequences so just don't go there!

Will any of the role players be pretending to have disabilities?

Yes quite possibly. Patients from all walks of life are included in this exam so don't be surprised if you are faced with a role player who is pretending to be blind, deaf, or have learning difficulties. Think about how you manage these consultations in everyday practice and if you have never had to think about this before, then practise with your colleagues. For example, it's always best to establish early on in the consultation how a deaf person would like to communicate with you, e.g. lip reading/writing things down. If you don't do this then you run the risk of ignoring their disability and either carrying on regardless or shouting so loudly at the patient that it's just embarrassing for everyone involved.

> Tip: Think about how you would manage a patient with a disability such as deafness and remember to establish how they would like to communicate with you early on in the consultation.

What's the best type of preparation for the exam in terms of revision?

Practice, practice, and practice—and this means seeing plenty of patients in your surgeries. Ideally at this stage you should be seeing between 90 and 100 patients per week. Make sure that you are regularly working to at least 15 (if not 10) minute appointments and that you are running on time. Granted, in the exam you will not need to write notes or fiddle with the computer screen, but you may need to allow time to examine the patient and should be doing the majority of the consultation within 10 minutes.

> 'The best thing that I did in preparing for the exam was to make sure that I had been doing 10-minute appointments in my practice for a considerable amount of time. This was a good way of practising my time management skills.'
>
> ST3

Forming small study groups with your peers is a great way of practising not only your communication skills but also your timing and getting used to being observed consulting. Take it in turns to play the part of the doctor and the patient and have an additional person observing to make comments on what went well and how you could improve.

> Tip: Practice, practice, and practice—and this means seeing plenty of patients in your surgeries.

Tip: Take it in turns to play the part of the doctor and the patient and have an additional person observing to make comments on what went well and how you could improve.

During your practice role play with your peers, think about some of the more difficult opening statements made by patients that you may not have come across before and how you would respond to them. For example, think about some of the following:

- 'I'm pregnant, doctor'
- 'My husband has left me'
- 'I was raped last night'
- 'My 10-year-old daughter died yesterday'

Discussing bad news is never easy in any consultation, but make sure before you start pouring out the sympathy that what you **perceive** to be bad news actually **is** bad news for the patient. A 17-year-old who comes in saying she is pregnant isn't necessarily devastated and wanting an abortion, in the same way that some women if they have been physically abused by their husband for years may be delighted that their husband has left them. Some opening liners, however, are completely devastating and it's important to think about how you would respond and further manage the consultation.

Tip: Think about some of the more difficult opening statements made by patients that you may not have come across before and how you would respond to them.

There are several books out there which can help to focus the learning during these study sessions or you could even consider writing some cases of your own. This has the added benefit of putting yourself in the position of the examiner and thinking about what competencies they are trying to assess in each case.

Books which I can recommend to use during the preparation period include:

- **The 'Blue Book':** ***Cases and Concepts for the new*** **MRCGP (P. Naidoo, Scion Publishing Ltd.)** This book contains general information on the exam plus 42 scenarios which you can use to practise within your study groups. It gives notes for the actor and the doctor and also information for the observer to think about when assessing how the consultation went. There is an additional section at the end on case-based discussion and some other GP concepts that you may comes across in your everyday work.

- **The 'Red Book': CSA *Scenarios for the new* MRCGP (Thomas Das, Scion Publishing Ltd.)** This book is more concise and looks at common presentations that you may encounter in the exam and breaks them down into the three domains on which you will be assessed. It also provides a brief revision of red flags to consider in each case and how to incorporate the patient's ideas, concerns, and expectations in to the consultation. Working through this book is a good way of covering the common medical conditions and presentations and gives a good structure to learning.

Both of these books are available via the RCGP book shop online where as members you should receive a 10% discount. There are plenty of other useful revision books out there but don't waste too much money on reading books for this exam. Chose one or two books which you can use to focus your learning and leave it at that.

Tip: Chose one or two books which you can use to focus your learning and leave it at that.

Make sure that you have up-to-date knowledge of how to manage common conditions and a broad overview of some of the more frequently used NICE guidelines. Knowing the doses of commonly used medications and their interactions is also important although you are allowed to take a *BNF* in with you. To be honest though, frantically trying to find the right page in the *BNF* can waste a great deal of time and you shouldn't be expected to prescribe anything weird and wonderful with complicated dosing.

If you aren't able to form study groups or want additional practice, then videoing your consultations is also a good way of polishing your skills. Watching yourself consult in the comfort of your own home or even with your trainer, helps you to recognize the areas of your performance that could be improved and what gaps you may have in your knowledge base.

Tip: Videoing your consultations and watching them back is also a good way of polishing your skills.

Think also about what examinations you may be asked to perform and either video yourself doing these on real-life patients or practise them with friends. Making your examination technique look slick and proficient is important and is an easy way of scoring good marks. Don't worry about rehearsing those full systems examinations that you were expected to do in your medical finals though, as instead you will be expected to do a focused and relevant examination that is more appropriate to general practice. (See later question **'WILL I HAVE TO EXAMINE THE PATIENTS?'**)

Is there any particular consultation model that we should be using in CSA?

In terms of Pendleton versus Neighbour and all that jazz, then not really. What you need to do instead is focus on the three domains against which you are being marked and make sure that you remember to incorporate all three into the consultation. However, if you had to pick one to base your consultation on then Neighbour is probably the best, as it makes sure that you also think about safety netting, follow-up, and housekeeping, which not only means that you should get to the end of the consultation each time, but that you also don't collapse in a hysterical mess between cases if things go wrong!

> Tip: Practise ways of finding out people's ideas, concerns, and expectations without using clunky or cliché phrases. Often, finding out the patient's health beliefs early on makes it easier to extract this information in a more natural way than having to ask 'what were you hoping I could do for you today?'.

Should I go on a course?

This very much depends on how much study leave you are entitled to and whether you can afford it. Some may argue, however, that you can't afford not to—even if it means paying for a course out of your own pocket with no study leave reimbursement and having to take the time off as annual leave rather than study leave.

There are many courses running across the UK that are either locally organized or centrally organized via the RCGP. Most of the people in our cohort attended two courses—one being the RCGP organized course run at the examination centre in Croydon and the other being a locally organized course run by our VTS course organizer. Both were invaluable in helping us on our way towards CSA success with our local run course currently boasting a 100% first time pass rate.

The RCGP course that is held at No 1 Croydon is officially accredited by the college and is a fantastic opportunity to have a practice run at getting to Croydon, make yourself aware of the building, familiarize yourself with the environment, get to practise consultations in the same rooms that will be used in the examination, and also get to meet registrars from other VTS groups to exchange hints, tips, and general banter. It is a two-day course and there is capacity for around 50 delegates. The course is run by RCGP examiners and case writers, so you really are hearing everything from the horse's mouth as it were. There is also an opportunity to purchase exam revision guides at discounted prices if you so wish, although depending on when you attend the course it may be too late for all of that. At last check this course was

£450 to non-members and £330 to members which is an absolute bargain considering what you learn and gain in just two days. You are well looked after, very well fed, and will hopefully come away feeling much more prepared for the exam that looms ahead.

The locally run course which we attended was lovingly created and set up by our VTS course organizer, who recognized that perhaps not everyone can either afford to go to an externally run course or is able to spend a night away from home in London. She also realized that getting all of the trainees through the exam first time was much more cost-effective than having to apply for extended training for them if they failed. This course was run every Thursday afternoon for three consecutive weeks and between seven and ten trainees attended. We discussed the format of the exam, focused on what made good consultation skills, wrote our own cases in order to help understand what the examiners are looking for, and got to role play five consecutive cases on two separate occasions under exam conditions assessed by a combination of local GP trainers and real college examiners. Fortunately this course was provided free of charge after application for funding from our local deanery and has been a huge success. It now runs several times a year in the run up to the CSA exam dates and has been an invaluable addition to our core VTS teaching. If you don't have anything similar to this in your local area it may well be worth approaching your course organizers to see whether this is something that would be feasible for ST3s within your VTS. We certainly thoroughly enjoyed it and felt that it gave us a good head start when preparing for the exam.

There are also several other courses run in different locations across the country of varying length and cost. For more details of these as well as the accredited RCGP courses see the 'Courses & Events' section of the RCGP website.

Tip: If you don't have a locally organized course in your area it may well be worth approaching your course organizers to see whether this is something that would be feasible for ST3s within your VTS.

'My biggest piece of advice would be that you must go on a course beforehand to get a flavour of the CSA and the diverse case mix and time pressures that come in to play during the exam.'

ST3

When is the best time to go on a course?

This depends on course timings and availability, but try to go on a course as soon as you can once you start your ST3 year. It is not only important in terms of preparing you for the exam but getting yourself in to good habits in terms of consultation and communication skills from day one will make the rest of your ST3 year (and GP career for that matter) run more smoothly.

Should I have some time off for study leave prior to the exam?

This really depends on what type of person you are and how you normally prepare for exams. Unlike AKT, CSA is not the type of exam that last-minute cramming will help with and therefore you are probably better off continuing to work within your surgery up until the day of your exam, as seeing patients is the best way of practising for it. You may, however, consider having the day off before your exam, to give you plenty of time to get down to London without rushing or stressing. This is particularly advantageous if you are sitting the exam in a morning session.

If you do decide to have study leave prior to the exam, make sure that during that time you are still practising your consultation skills and aren't sat for hours with your head in books. Either being in surgery seeing real patients, or role playing case scenarios with your peers is the best way to spend the few days prior to the exam, so that you remain in 'consultation mode' and not 'book mode'.

> Tip: If you do decide to have some study leave prior to the exam, make sure that during that time you are still practising your consultation skills and aren't sat for hours with your head in books.

> 'I sat my exam on a Thursday and decided to work the Monday and Tuesday of that week so that I was continuing to practise my consultation skills almost up until the day of the exam. I decided to have the Wednesday as annual leave so that I could get myself prepared both practically and mentally and make my way down to Croydon in a leisurely fashion rather than rushing after work.'
>
> ST3

Can I cancel if I change my mind?

In short, yes—but only if you cancel within the application period designated by the college (as published on the website). As you will have already paid by this point, all fees paid will be transferred to the subsequent sitting that you decide to apply for. If for some reason you do not intend to ever sit the CSA exam in the future, then you can apply for a refund via the college. If, however, you chose to cancel your exam outside of the designated period on the website then you will normally have to forfeit your assessment fee. The same is true if you do not turn up to the examination on the day. If you feel that there were extenuating circumstances which required you to cancel or not attend the exam, then you may apply for a refund by writing to the college and explaining these circumstances in full. If this is the case you may be asked to provide evidence of why.

How do I get to Croydon?

Probably the easiest and least stressful way of getting there is by train, as London traffic is awful at the best of times and parking at the centre is limited only to disabled spaces.

- By train—frequent fast trains run from Victoria and London Bridge so you need to aim to get to these destinations within central London as your first hurdle. The journey time from Victoria is about 15–20 minutes and once you reach East Croydon Station, the rest is on foot.
- By car—probably the easiest thing to do would be to set your sat nav to take you to East Croydon Station (or use the postcode CR0 0XT) and if you don't own one of these expensive navigation devices then look up the route to the station via an alternative trusty source such as AA route finder (http://www.aa.com). You could also use a good old-fashioned map if you like that sort of thing. Essentially if you get somewhere close you should be able to follow the road signs to East Croydon Station but it all really depends on where you are planning to park. If you are not eligible for a disabled parking space at the centre then you will need to park at one of the alternative car parks that can be found within half a mile of the NLA Tower.
- By bus/tram—perhaps a more difficult and stressful option but plenty of buses and trams stop at East Croydon Station. Check out http://www.tfl.gov.uk (transport for London) for further information.

Should I stay overnight?

Staying in a hotel or in other accommodation the night before the exam is definitely sensible if you have chosen to sit the exam in a morning session. Rush hour in central London is hectic enough without first having to get to London from your home destination and then make your way across to Croydon. Think in advance about how long it will realistically take you to get there allowing for train delays and traffic and see what you think. Whilst a night in a hotel feels like an extra expense, at least you can relax in the knowledge that all you have to rely on is your own two feet to get you to the exam on time. For afternoon sessions it may be less crucial to stay over the night before, but again it depends on how far you have to travel and what delays you may encounter along the way.

> Tip: If you are coming from afar then staying in a hotel or in other accommodation the night before the exam is definitely sensible if you have chosen to sit the exam in a morning session.

Where can I stay overnight?

The closest four hotels are detailed in Table 18.2 and are all of varying price range depending on when you book and what type of room you need.

Table 18.2 The four closest hotels to the NLA Tower

Hotel name	Distance from station (km)
Croydon Park Hotel	0.1
Croydon Central Travel Lodge	0.3
Jury's Inn	0.4
Express by Holiday Inn	0.5

What should I wear?

Something smart but comfortable. That doesn't necessarily mean a suit, although the males amongst you really should wear a shirt and tie at least (and trousers obviously!). For females the skirt/trouser debate isn't really worth worrying about, but obviously don't turn up in a mini skirt, fishnet tights, and a cleavage-showing top. Whilst there are possibly examiners out there who would appreciate this approach, remember that this is after all a professional membership exam and it should be treated as such. You do, however, need to be comfortable, as you are there for three to four hours and if you don't normally wear a suit to work you may feel strange doing so for the exam and it may affect your consultation style.

What do I need to take with me on the day?

Two very important things:

1. **Doctor's bag** with the following contents:
    - BNF—*must not contain any handwritten notes and may be checked before the start of the assessment*
    - *Stethoscope*
    - *Ophthalmoscope*
    - *Auroscope*
    - *Thermometer*
    - *Patella hammer*
    - *Sphygmomanometer (manual or electronic)*

- *Tape measure*
- *Peak flow and disposable mouthpieces (EU standard)*

2 **Photographic identification documents**—valid passport or photo-card driving licence. Nothing else will be accepted. Make sure that your first and last names match your identity documents **exactly** otherwise you may run in to problems (unless you bring your original marriage certificate for example).

You don't need to take a stopwatch as each consultation room is equipped with a large clock, but if you'd rather have something close at hand then that's fine. The chances are you will be so involved in the cases that you probably won't even look at it, but if it makes you feel better then go for it.

Other things you may wish to take include a copy of your confirmation email and a map of the local area.

On the day

What happens when we arrive at the centre?

You should arrive at the centre by 9.00am for a morning session and 12.30pm for an afternoon session, to allow time for checking in. See Fig. 18.1 for the order of events on arrival.

- No candidates will be permitted to complete the assessment if they arrive after the briefing has started
- Signing of the non-disclosure agreement ensures that you agree to not pass on any knowledge of any of the cases, as the assessment material is confidential and copyrighted to the RCGP
- Three identical circuits of consultations will run simultaneously over the three floors, so it's important that you know where you are going
- Mobile phones must be switched off and left in your locker. If for some reason you may need to be contacted in an emergency then you can leave your mobile phone with a marshal, who will answer it for you and contact you if necessary
- The exam will not start until all candidates are settled in their rooms and any necessary checks have been completed. A bit like that bit when you're on a plane just before take-off and equally as scary!

How long does the exam last?

The actual exam lasts about three hours but you'll probably be at the centre for around four hours in total. For a morning session you will be required to be there by 9.30am at the latest as this is when the candidate briefing will start. Following this you will

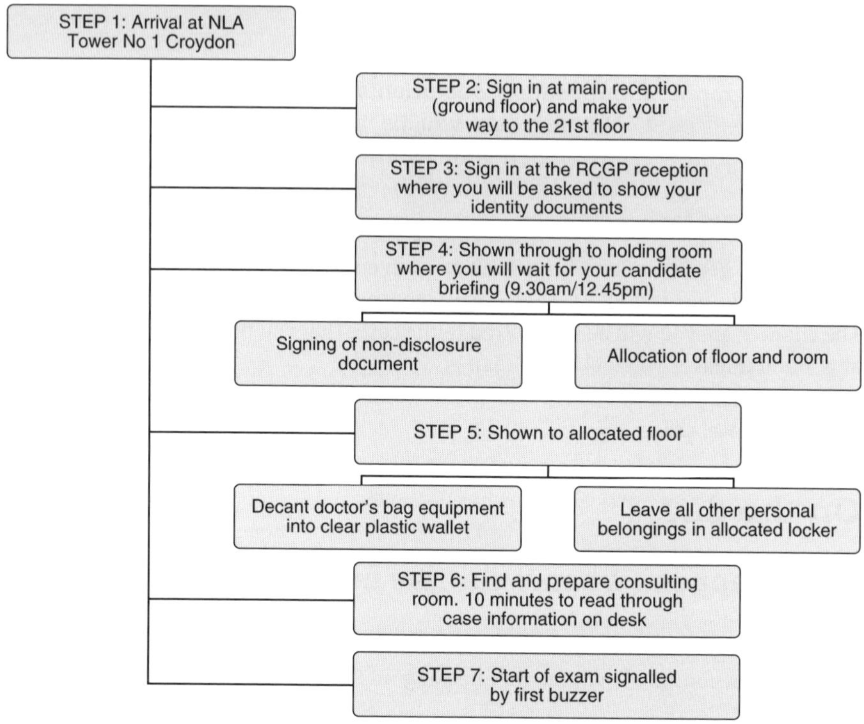

Fig. 18.1 Order of events for CSA on arrival at the assessment centre

be taken to your consultation room where you will remain for the next 162 minutes completing the exam. You will then be quarantined until 1pm when all of the afternoon candidates have been taken into the briefing room in preparation for the afternoon session. The same will apply in the afternoon but you will be allowed to leave as soon as the exam finishes as there will be no need for any quarantining. Afternoon candidates need to be at the centre by 12.30pm and for the security of assessments will not be admitted if they arrive any later than this. Make sure that you carefully check your confirmation email for further details and to check you have booked for the right date and time. Candidates who arrive after the start of the briefing for either session will not be admitted to the assessment under any circumstances.

What will I find in my room?

On entering your room you will find a desk, a chair, a clock, mock FP10s, and a pile of case notes in the order of the patients that you are going to see (for further information see the next question **'HOW MUCH BACKGROUND INFORMATION WILL WE BE GIVEN FOR EACH CASE?'**). You will also be provided with a small selection of extra equipment that may be required for some of the cases. You certainly do not have to use this equipment (so don't panic if you don't use all of the items) but if something is required that isn't on the 'list of things to bring in your doctor's bag' then it will be there for you. And just a tip on bringing your doctor's bag—don't forget anything! Because if you do, there will not be spare supplies and you may end up doing something daft like sticking your biro in someone's ear by way of pretending that it is an otoscope! This may lead to you failing the case.

There will also be an air conditioning control unit near the door which you are allowed to fiddle with if you feel brave enough. Many people found the centre to be quite cold in general because of the air conditioning, but after a few cases and some adrenaline surges you soon heat up and will be glad of the cold breeze!

> 'When we were first shown in to our rooms I couldn't get over how cold it was! But then I realized that maybe I was just shivering with nerves, as by the end of the first half of the exam I felt sweaty and wished I knew how to turn the air con up!'
>
> ST3

Tip: Don't forget any of the essential items for your doctor's bag. If you do, there will not be spare supplies and you may end up doing something daft like sticking your biro in someone's ear by way of pretending that it is an otoscope! This may lead to you failing the case.

How much background information will we be given for each case?

On the desk you will find a list of the patients that you will see during the surgery along with a summary sheet for each case, in the order that they will be seen. Each summary sheet will contain background information which you may find useful during the consultation. This may include relevant past medical history, drug history, notes from a previous consultation, or information about the patient's social habits. You are able to make notes on these sheets but they must not be removed from the room at the end of the examination. Make sure that you read each case thoroughly as some cases may be spread over two or more pages and contain additional pieces of information such as test results or hospital letters. You will not have a computer in your room and you will not be expected to do any computerized note keeping or submit any handwritten notes based on your consultation. This may, however, change in the future as note keeping and litigation become even more important issues to consider.

How long will I have for each case?

Ten minutes. The start and end of which will be marked by a buzzer sound which they test to check that everyone can hear before the exam starts. The role player and examiner will also get up and leave when the end buzzer sounds, so if you do happen to miss it you'll soon find out that the case is over. Nothing that you say or do after the 10 minutes is up will be counted towards your mark. After the examiner and role player have left the room you then have two minutes to read through the notes for the next case and do any preparation that is required. The buzzer will then sound again to signal the start of the next case and so on.

How many cases will I see? Will they all count towards my final mark?

Every candidate will see 13 cases, each of 10 minutes' duration which all count equally towards your final mark. Between each case there will be a two-minute break for you to compose yourself and prepare for the next case.

> 'I nearly lost the plot after one of my cases had gone really badly but managed to sort myself out by remembering that I didn't need to pass them all and that no one is perfect—we all make mistakes. It made me really realize the importance of 'housekeeping' after the consultation!'
>
> *ST3*

Who will be examining me?

All of the cases are assessed by MRCGP examiners who have been trained in assessing postgraduate general practice. The examiner follows the role player throughout the session so that they mark the same case all day. This enables the cases to be well calibrated and promotes validity and reliability of the results. The examiners will follow the role player into the room when you call them in and should sit outside of the candidate's line of view. They will not interact with you unless required to do so (for example, to stop you performing an unnecessary examination or to give you the findings of an examination that you have said that you would like to perform).

It's worth mentioning also that some of the cases will be videoed by a discreet video camera set up within the room. Again this is for the purposes of the college training or validation and is not optional (e.g. you can't refuse to be videoed even if you haven't got your best suit on!) You probably won't even notice anyway.

How many examiners will there be?

In the vast majority of cases, each role player will be accompanied by one examiner. This examiner stays with the role player for the whole morning so effectively gets to see 13 different candidates approach the same case. You can imagine they must get pretty sick of this by the end of the session and it is pretty stressful and tiring for them too.

There may, however, be circumstances where new examiners are being assessed or validated so that you may have two official people enter the room. Try not to be put off by this as the second person will not be assessing you and in fact you may be at an advantage by the examiner having to be very conscientious about the way in which they mark you for fear of getting it wrong themselves!

Will I have the same examiner for each case?

No. The examiner and the role player stay together and come to you in your consulting room, so you will be assessed by 13 different examiners. This makes the examination much more valid and reliable and means that if you really mess up a case then you don't have to worry about it tarnishing the examiner's view of your performance overall.

Will I need to examine the patients?

Yes, yes, and yes! But only if appropriate. Obviously don't offer to do a vaginal examination on someone with a headache and there are few circumstances in which you should need to do a full neurological examination. But you **must, must, must** assume that you are required to carry out an examination if the consultation requires it. Looking anxiously at the examiner for a response when you say that you would like to examine the patient will just lead to embarrassment and flustering as they stare back at you blankly.

If you need to do an examination then explain this to the patient as you would do in a normal consultation and proceed to do so (but don't forget to ask their permission beforehand!). If you are not required to do this the examiner will stop you and give the examination findings. Although, bear in mind that they will only give you findings of those examinations which you have requested to perform on the patient, so make sure that you make it clear **exactly** what you would like to examine including things like blood pressure or pregnancy tests. If you don't ask, you don't get and that can make a huge difference in the case of a young female with abdominal pain in whom you forget to ask for a pregnancy test! However, also be prepared for them not to stop you from examining the patient, as in about one-third of the cases you will be allowed to proceed to examine the patient. When doing so make sure that what you are doing is relevant, focused, and done with proficiency. Tapping out for ascites in someone who has come with simple dyspepsia and has a very obviously normal abdomen will just look ridiculous. You could also get given a model to examine, e.g. breast or prostate, so try not to look too shocked when the examiner produces a plastic organ from his pocket for you to demonstrate your examination skills on!

> Tip: When examining the patient, make sure that what you are doing is relevant, focused, and carried out with proficiency.

In your preparation, it's therefore worth running through common targeted examinations that you may be required to perform in order to look slick and competent on the day. For example, how to examine a thyroid, swollen leg, abdomen, diabetic foot, etc., and how to do it well. Unlike MRCP PACES, this exam is not really a test of your examination skills, so any examination you do shouldn't take up too much of the valuable 10 minutes, but should be enough to prove that you know what you are doing and why you are doing it.

WILL THE PATIENTS HAVE ANY CLINICAL FINDINGS ON EXAMINATION?

If you mean will you find a murmur when you listen to their heart then maybe, but it would be extremely unusual. All role players and their cases are thoroughly scrutinized on the day of the exam by the examiners so that the case is played identically by the role players on all three floors. If an examination is required as part of the case then the examiner will have performed this examination on the role player to discover whether there are any abnormal findings that you may come across. Again, this isn't MRCP paces so you are unlikely to come across this scenario but don't be baffled if you do find some positive clinical findings. There is talk at the college that over the coming years they may well change to having role players with clinical findings in order to further assess your clinical skills but that certainly isn't really the case at the moment. As you can imagine, it would be very difficult for the college to find three actors (one for each floor) who all had a systolic murmur or a seborrhoeic keratosis on their back, however, there are some clinical findings which could be role played, e.g. visual field defects or tremor, so this may be something for the future. If you were to find a murmur on a role player that wasn't supposed to be there, I'm sure that the examiner would correct you before you launch into a major discussion about organizing an echo, etc. for someone who came in with a chest infection (but in whom you just happened to stumble upon their benign systolic murmur whilst examining their chest!).

Will I need to write any prescriptions or certificates?

This depends on the case, but if you feel that a prescription or fit note is required, then you have two options. You can either explain exactly what you would prescribe or write on the certificate and say that it can be collected from reception, **or**, you can actually go ahead and write them on the mock FP10s or fit notes provided. But remember, anything that you give to the patient is open to being assessed by the examiner and therefore think carefully before hurriedly scribbling out a prescription that is either incomplete or incorrect. As long as you explain to the patient exactly what you intend to do then you should be fine.

What if I want to organize any tests/investigations?

Again, if this is the case, you must fully explain to the patient what tests you would like to perform and why (plus any special instructions, for example, to fast for 12 hours before a glucose tolerance test) and when you would expect to receive the results. You do not have to give the patient actual forms to take away with them for these tests, as lab/x-ray forms are likely to differ hugely across the country (that and watching you fill out a tedious form is not a good use of the CSA time). This may, however, change in the future so make sure you listen out for any alterations as to what is and isn't expected of you within the exam.

Can I give the patient leaflets?

Without a computer or printer you obviously can't physically give the patient a leaflet, but there will be scrap paper which you can use to write down essential information or websites for patients. Try, however, to avoid offering everyone a pretend leaflet that they can pick up from reception, unless you have discussed with the patient what information it will contain. It is not a substitute for explaining something properly to a patient and becomes an empty gesture if it isn't really relevant.

Should I wash my hands between patients?

There won't actually be any facilities for you to wash your hands within your consulting room but you should ideally indicate to the patient that you would wash your hands before/after examining them, depending on what your normal practice is. Unlike in your driving test though, don't expect the examiner to give you a big congratulatory smile when you suggest that you would do this (like they would if you did a very obvious 'mirror, signal, manoeuvre' before driving away from the test station) as most of them probably won't even notice. However, it's worth just making it clear that this would be your usual practice.

Do we get a break during the exam? If so is it at the same time for everyone?

Yes and yes. After the first seven cases there will be a 15-minute break where you are all ushered out into a central area where you can nervously slurp a coffee, munch on a HobNob, or be escorted to the toilet by an invigilator. You are allowed to chat with the other candidates but it obviously cannot be exam-related chat and all conversation is closely monitored by the staff.

Will there be any children?

The RCGP at present are not actually allowed to use any child actors, but there are plenty of adults out there who could easily pass for young adults or teenagers. Cases that involve young children may, however, be included in the exam but will, for example, be in the form of a parent coming to discuss a problem that their child is having or a health visitor wanting to discuss a case with you. It's therefore essential that you feel confident dealing with common childhood complaints and conditions. It's also worth brushing up on the GMC confidentiality in children guidelines, as you may well get a parent trying to find out confidential information about their Fraser competent 15-year-old who doesn't want their parents to know what they've been up to.

Tip: Brush up on the GMC confidentiality in children guidelines as you may well get a parent trying to find out confidential information about their Fraser competent 15-year-old who doesn't want their parents to know what they've been up to!

Will I have to do home visits/telephone calls?

Yes, you may be expected to do these. If so, you will be collected from your room by the examiner and taken to a separate room where you will either find a role payer pretending to be in their own home or a telephone point with two sets of headphones for both you and the assessor. Try not to be fazed by either of these scenarios and just proceed as you would do normally. In actual fact, depending on where you work it may be very far from what you would experience normally, as there won't be any mouldy food on the floor and you won't have to navigate a keypad lock to get into the room! You also shouldn't get attacked by any mangy dogs or have to crouch next to the patient in fear of sitting on their urine stained couch... but you never know!

Tip: If you are faced with a telephone call as one of your cases, always make sure that you not only introduce yourself but that you check the identity of the caller and bear in mind confidentiality issues.

What if the 10-minute buzzer goes before I have finished the case?

Tough luck I'm afraid. Nothing that you say or do after the 'end of case' buzzer has sounded is marked, which is why it is so important to make sure that you are practised at consulting in 10 minutes and that you try and give every consultation a beginning, a middle, and an end. Failure to get to the management section of the case may well mean that you miss the whole 'nub' of the case and will therefore gain fewer marks. If you're someone that frequently tends to go over time, make sure you keep an eye on the clock and if you're getting close to the end of the 10 minutes, move on to a management plan even if you feel there is still more data gathering that you would have liked to do. You can always suggest that the patient makes another appointment to discuss things further or say that you need to get some further advice from a colleague and will phone them later if you are stuck.

Tip: If you're someone that frequently tends to go over time, then make sure you keep an eye on the clock and if you're getting close to the end of the 10 minutes, move on to a management plan even if you feel there is still more data gathering that you would have liked to do. You can always suggest that the patient makes another appointment to discuss things further or say that you need to get some further advice from a colleague and will phone them later if you are stuck.

Tip: If not finishing the consultation in 10 minutes is your main concern then make sure that your exam preparation involves videoing lots of COTs, keeping a watch on your desk during your normal surgery, and making sure that you are starting management negotiations by at least the eight-minute point. (RCGP Examiner)

What if I finish the case before the 10 minutes is up?

Firstly, don't be alarmed by this if it happens. Some cases just are more straightforward than others so try not to worry that you have missed something obvious (unless you finish **very** early in which case there may well have been something else to have uncovered!). Remember, the cases are designed so that you should be able to complete them in no more than 10 minutes, so if you finish a case early, enjoy the extra few minutes of relaxation and move on to the next case as normal. It's also worth mentioning that not all patients will leave the consultation satisfied (for example, the lorry driver wanting a sick note to cover the amount of time for which he has been banned for drink driving in order to prevent him losing his job) and if someone gets up and walks out in a huff after eight minutes, it doesn't necessarily mean that you have done anything wrong or that the case has gone badly.

How is the exam marked?

All 13 cases are assessed by a trained RCGP examiner using a pre-set marking schedule specific to the case. Three domains are assessed and a mark is given for each. From this, an overall numerical mark is given for the case.

The three domains are as follows:

- **Data gathering, technical, and assessment skills**

  This looks at how you gather information in the form of taking a history, examining the patient, and the relevance and analysis of any investigations. Your proficiency in examination skills is also assessed (see **'WILL I NEED TO EXAMINE THE PATIENTS?'**).

- **Clinical management skills**

  This area looks at your ability to recognize and manage common medical conditions and demonstrate your ability to make decisions based on your findings. It also assesses how you deal with multiple complaints and comorbidity.
- **Interpersonal skills**

  This domain concentrates on how you communicate with the patient and come to a shared management plan by understanding their health beliefs and recognizing relevant equality and diversity issues.

The number of marks for each case is then added together to create a final mark for the exam. As well as assigning numerical marks to each case, the assessor will also rate each candidate as a pass, fail, or borderline overall. For each case the overall numerical case marks of the candidates in the borderline group are then averaged. These averaged scores are then aggregated across all the 13 cases to create the pass mark for that case which then differentiates a pass from a fail. Using this method means that the pass mark for each day of the exam will be different.

This change was created in 2010 after the previous system of using pass, borderline pass, borderline fail, and fail for each case, did not discriminate well between the good and bad candidates. The revised method also allows the pass mark to reflect the difficulty of the cases attempted that day.

Special circumstances

What if I am ill?

If you are feeling under the weather but do not have a serious illness then my advice would be to try and attend and give it a go. Unless there are significant extenuating circumstances then you are unlikely to get your fee refunded and you never know, the adrenaline might just get you through on the day! If, however, you really are acutely unwell and not able to attend then you may apply to the college for reimbursement of your fee, which will be completely at their discretion. My advice would be to at least let the college know in advance if you possibly can, as they are then likely to look kinder on you when you apply for your refund (see earlier question **'CAN I CANCEL IF I CHANGE MY MIND?'**).

> Tip: If you are unable to attend the exam due to illness then try to let the college know in advance as they are then likely to look kinder on you when you apply for your refund.

Do you get special consideration if you are pregnant?

In short the answer is 'no' but if you inform the invigilator on the day or in advance, then you may be lucky enough to get to have your room as close as possible to the toilet. There is no extra time given to women who are pregnant and if they require the use of the toilet during the exam then they must do this either during their two-minute break between cases or if they finish a case early. If you need the toilet outside of these times, unfortunately you will forfeit your time with the simulated patient. Whilst this seems unfair, it is essential that the exam runs to schedule and these types of allowances would cause havoc for the remaining candidates and examiners.

> Tip: If you will be pregnant on the day of the exam, let the college know in advance so that they can try and place you in a room as close as possible to the toilet for those all essential toilet breaks!

Do the college have a policy for those with disabilities?

Yes. Candidates with a medical condition or disability which may affect their capacity to undertake the CSA should ideally have notified the college at the time of registering for the exam. The RCGP states that they make 'every effort to ensure that adjustments are appropriate, proportionate and consistent and that they do not result in an unfair advantage'[2].

In order to assess what type of adjustment is appropriate for your specific disability or medical condition, they may seek further information about special arrangements that have been made or were required during other aspects of your training or assessments. For full guidance on their policy for those candidates with a disability there is a full paragraph within the document *nMRCGP Clinical Skills Assessment – Information for Candidates*[2] on the RCGP website.

What if it snows?

This may seem to some like a strange question, but in January 2009 there were serious problems with snow affecting the smooth running of the winter examinations. With actors, assessors, and candidates all struggling to make the fretful journey into London because of the heavy snow, there were a few days of exams that had to be abandoned and rescheduled for a later date. I therefore cannot express enough how important it is to keep a close eye on the RCGP website in the weeks running up to the exam just to make sure that everything still appears to running smoothly. It may also be worth leaving booking trains and hotels until the last minute should there be any reason to think the exam may not proceed as planned, as you will be unable to get reimbursement from the college should you have to cancel these things. If, however, you are nervous about leaving things until the last minute then I would suggest to just go ahead and book and worry about the money lost at a later date. It could be even more stressful if you find when you come to organize your accommodation that all the local hotels are full and you end up having to stay somewhere miles away from the exam centre.

Tip: Keep a close eye on the RCGP website in the weeks running up to the exam just to make sure that everything still appears to running smoothly.

To what extent could something like the 'swine flu outbreak' affect CSA?

The college currently has a page on its website designated to providing information on this. Essentially all exams will plan to run as scheduled but should there be a pandemic, your exam may well be cancelled and your fee transferred to the next possible sitting. If you should be affected by this and require further information then you can contact the RCGP examinations department on 020 7344 3212 or via email at exams@rcgp.org.uk.

Who can I contact in an emergency on the day?

The RCGP assessment centre staff can be contacted on 0208 253 4353/5.

After the exam

Where can I get some food?

There are no catering facilities within the building but there are plenty of places to get food if you head back towards the station or into the town centre in Croydon. Taking a snack and a bottle of water with you is not a bad idea as you will no doubt feel pretty hungry and thirsty after the exam and it may be a while before you find a suitable eatery in which to park your bum and get some sustenance.

> 'I always seem to get horrendous headaches after any exam so I made sure that I'd packed a good supply of paracetamol, ibuprofen, and water before I left the house. It was amusing how many people were asking to steal some in the lift on the way down. That's good old tension headache for you!'
>
> ST3

How do we find out the results?

The results can be found on your ePortfolio in the section 'Progress to Certification' and are not published elsewhere on the RCGP website. If you have passed you will find a very satisfying big green tick next to the CSA, with an icon on which you can click to reveal your assessment breakdown and feedback.

When do we get the results?

The results appear on your ePortfolio anything from one to three weeks after the last exam has taken place in that sitting. The exact dates can be found on the RCGP website and really depend on how many people sit the exam and the time of year. If you're someone that will lie awake at night worrying yourself silly about things you may have done wrong, then perhaps chose a date towards the end of the sitting so that you don't have to wait too long to find out the results.

> Tip: If you're someone that will lie awake at night worrying yourself silly about things you may have done wrong, then perhaps chose a date towards the end of the sitting so that you don't have to wait too long to find out the results.

Do we get any feedback on our performance in the exam?

Yes. This can be found by clicking on the icon next to the green tick or red cross adjacent to the CSA part of 'Progress to Certification' on your e-portfolio.

The first part of the feedback gives a breakdown of the cases that you have seen by curriculum statement heading, with your score beside each one as either pass, borderline, or fail plus the numerical mark that you were given for that case.

The second part of the feedback is given for the assessment overall, rather than for each consultation and provides 16 statements which suggest areas in which the candidate's performance was deficient. If any of these statements are identified by two or more assessors then it will be indicated by an X next to the relevant statement.

The statements are as follows:

Data gathering:

1. Disorganized and unsystematic in gathering information from the history taking, examination, and investigation
2. Does not identify abnormal findings or results or fails to recognize their implications

3. Data gathering does not appear to be guided by the probabilities of disease
4. Does not undertake physical examination competently, or use instruments proficiently

Clinical management:

5. Does not make appropriate diagnosis
6. Does not develop a management plan (including prescribing and referral) that is appropriate and in line with current best practice
7. Follow-up arrangements and safety netting are inadequate
8. Does not demonstrate an awareness of management of risk and health promotion

Interpersonal skills:

9. Does not identify patient's agenda, health beliefs, and preferences/does not make use of verbal and non-verbal cues
10. Does not develop a shared management plan or clarify the roles of doctor and patient
11. Does not use explanations that are relevant and understandable to the patient
12. Does not show sensitivity for the patient's feelings in all aspects of the consultation including physical examination

Global:

13. Disorganized/unstructured consultation
14. Does not recognize the challenge (e.g. the patient's problem, ethical dilemma, etc.)
15. Shows poor time management
16. Shows inappropriate doctor-centeredness

This information provides formative feedback not only for the candidates who failed and will need to re-sit the exam, but also for those who passed but would benefit from feedback on how to further improve their performance.

If none of these areas are flagged by a 'X' then you may assume that these areas were identified as deficient by fewer than two assessors or perhaps none (and in which case then give yourself a well deserved pat on the back!).

How long is the CSA valid for?

From 1 August 2010, CSA passes are no longer subject to the three-year validity limit. Similarly if you passed the exam between August 2007 and July 2010 then your pass will remain valid pending the award of your CCT.

What if I fail? How many times can I re-sit?

From August 2010 a maximum of four attempts at the exam will be permitted. If, however, you started your training prior to August 2009 then you can have unlimited attempts at the exam as long as you retain your NTN.

Can I claim back my travel and accommodation expenses?

Yes, but you can only claim for reasonable public transport costs to attend the exam and the claim would need to be made via the PCT for whom you work, e.g. whilst you are in a GP training practice post. You are discouraged from taking any assessments during a hospital post and will therefore not be eligible to claim back these costs if this is the case. For those who are eligible to claim, you will be required to provide evidence in the form of receipts and can only claim for other reasonable expenses such as subsistence and accommodation if your exam is before midday.

Additional questions?

If after reading this chapter you still have some unanswered questions about the CSA, then scribble them down that you don't forget to ask your trainer or course organizer at your next VTS teaching session. The chances are that someone else will be pondering over the same thing and the only way to find out is to ask!

And finally... good luck!

REFERENCES

1 RCGP. *Clinical Skills Assessment*. Available at: http://www.rcgp-curriculum.org.uk/nmrcgp/csa.aspx (accessed 12 August 2010).

2 RCGP. *CSA Information for Candidates*. Available at: http://www.rcgp-curriculum.org.uk/nmrcgp/csa.aspx (accessed 12 August 2010).

CHAPTER 19

The job race!

What type of job?

As you near the end of your training you will be plagued by people asking you what type of job you are looking for, whether that be a partnership, a salaried post, working as a locum or something completely different. Well, the honest truth is that there aren't as many partnerships out there as there used to be and the majority of vacancies will be for salaried posts or locum work, but it's worth having a good old think about what it is that you really want in both the short, medium, and long term so that you can set yourself some goals and be prepared for the long road ahead. Don't be afraid to consider other options too, such as teaching, academic work, or clinical assistantships in hospitals. The beauty of having a general practice qualification is that you are a 'generalist' and are therefore able to work in many different fields of medicine without having to do too much additional training. This is where your special interests come into play and the true flexibility of general practice becomes apparent.

However, in terms of clinical general practice jobs, there are three main common types of work available and this chapter aims to provide a brief rundown of what each one involves and answer some of the common questions that may be asked about each type of work.

Partnerships

WHAT IS A PARTNERSHIP?

In short, a partnership consists of a group of people (normally GPs but sometimes practice managers and/or senior nurses) who own a GP practice and run it as their own business either in rented premises or in premises which they own. The number of people within the partnership varies and the number of patients is determined by the number of doctors in the practice. The responsibility for the practice is often divided between the people involved, in a way that reflects their clinical, financial, and administrative contributions. The yearly profit is then split accordingly to this partnership share. Being a partner is effectively like running your own business and means that you will be involved in all of the additional responsibilities that this brings. The partnership is bound by an agreement which is written by and agreed on by those within the partnership and therefore entitlement to study leave, maternity leave, sick leave, etc. will vary from practice to practice. As with any business there may be ups and downs in terms of profit and loss and therefore your salary will reflect how well or how badly the practice is doing. Some months/years you may get a better salary than others but in general, if you are working for a successful partnership your take home salary pro rata will be more than that paid to a salaried GP.

Think carefully about committing yourself to a partnership in a practice that you are not familiar with and even more so if it's within a completely new area. Starting work as a fully qualified GP is hard enough without having to learn a whole new system and area. Joining a partnership can be incredibly challenging and is not something which should be taken on lightly. Most partnerships will expect a period of mutual assessment before a newly appointed GP joins as a partner. During this time the GP in question will usually be salaried, with the mutual assessment time being anything from three months to two years. Once this mutual assessment period is complete, then you may become a 'fixed share' partner for the first year before becoming 'full profit sharing'.

> 'I am not sure if my experience of starting out in a partnership is particularly typical. It has been very challenging. I am far away from the area I did all my training in, and haven't got a good support network up here yet. My husband is rapidly becoming an expert in the ins and outs of general practice!'
>
> *Newly qualified GP after moving to a partnership in Scotland*

> Tip: Think carefully about committing to yourself to a partnership in a practice that you are not familiar with and even more so if it's within a completely new area. Starting work as a fully qualified GP is hard enough without having to learn a whole new system and area.

HOW MUCH DO GP PARTNERS GET PAID?

Answering this is almost as difficult as answering the 'how long is a piece of string?' question, but in general terms a GP partner will get paid more per session than a salaried GP. However, in view of the additional responsibilities involved you may or may not feel that the additional pay is worth it and that depends enormously on where you work and who with.

DO I HAVE TO BUY IN TO A PARTNERSHIP?

You should only have to buy in to a partnership if the partnership owns the premises from which they run. What you are then effectively 'buying in to' is the mortgage that they have on that property and the true assets of the practice, as opposed to 'goodwill'. By that I mean you are not buying their 'patients' as you would buy 'customers or market' if you were buying in to a commercial business. Whether you have to buy in or not also depends on whether or not there is an outgoing partner who wants their 'share' of the premises value. If this is the case then you may be asked to provide this amount of money to effectively take over the outgoing partner's 'share' of the business. Some partnerships, however, may choose to pay off an outgoing partner between the other existing partners, therefore not requiring you to 'buy in'. This of course would mean that whilst you will get your share of the partnership profits, you will not own any of the building and when you leave the partnership will not get a lump sum back. Not buying into the practice may also affect whether or not you are an equal partner.

HOW MUCH ANNUAL/SICK/MATERNITY LEAVE WILL I GET AS A PARTNER?

As a partner, your contract will depend entirely on what is written in the partnership agreement, which may well have been written many years ago (although all practices should have updated their partnership agreement when the new GMS/PMS contract came in. It should also be amended every time there is a change in the partnership). Make sure you check this in detail before agreeing to anything, as once you've signed it there'll be no going back. Unlike a salaried post which normally follows the BMA model contract, the conditions for a partner will vary hugely from practice to practice as they will essentially write the rules and regulations themselves.

Salaried posts

WHAT IS A SALARIED GP?

In basic terms, a salaried GP is one who works within a GP surgery for a specific amount of clinical sessions per week for an agreed salary. The amount of sessions worked varies enormously on what the practice needs and can be anything from one to ten sessions although few GPs work more than eight or nine sessions per week these days. Salaried posts sometimes also involve additional responsibilities which you may or may not be remunerated for in addition to the sessional rate, e.g. organizing QOF data or providing minor surgery sessions. Most salaried posts will also include paid sick leave, maternity leave, and annual leave although the amount that you are entitled to may vary from practice to practice. Practices that work under a GMS contract are obliged to offer a contract that is at least as fair as the standard BMA contract[1] which can be found on the BMA website. It's worth mentioning that the BMA also offer a service for members whereby any new contract that you are given at the start of a job will be looked through by their professional advisors to make sure that what is being offered is fair and acceptable. They will often suggest changes that should be made to the contract before you agree to sign it. Many practices will also offer to pay your GMC or medical defence organization payments although are unlikely to pay for both. Make sure that you factor this in when considering the sessional rate or salary that they are offering, as these subscriptions can be expensive once you're qualified and are doing plenty of clinical sessions.

WHAT IS THE 'MODEL CONTRACT'?

The model contract is a 'model offer letter' for salaried GPs, published by the BMA (but also agreed on by the NHS confederation, the Department of Health, and the GPC) that clearly sets out the 'minimum terms and conditions that should be used by a GMS practice or PCO when employing a salaried GP after 1st April 2004'[1]. It represents good employment practice and when it was produced it was hoped that it may also help those employed by PMS or APMS practices to negotiate improved terms and conditions in their contracts (as at present these practices are not obliged to follow the model contract).

The model contract applies to any GP working on a salaried basis for a GMS practice or PCO, whether that is as a standard salaried GP or a GP retainer, assistant, associate, returner, or a GP on a flexible career scheme. Their employment, however, must have commenced either on or after 1 April 2004. There are significant consequences for GMS practices that do not adhere to this model contract, to the point where they may find their GMS provider contract removed by their PCO. It is therefore suggested that all salaried GPs take advantage of the free BMA contract checking service for members before signing any agreements.

An example of the model contract which can be downloaded and completed by your appointing practice is available via the BMA website if you enter 'model contract' in the search box.

HOW MUCH DO SALARIED GPS GET PAID?

This varies enormously across the country and depends on how many sessions you do, the workload involved, and what else is included in your package. It also depends on whether you are expected to do additional admin, home visits, or on-call commitments but in general it ranges anywhere from £6,000 to £11,000 per session. If your medical defence payment is included in your contract then your sessional rate may be less, but in general the average salary is around £64,000 for an eight-session salaried post.

Another factor which comes into play is whether you work for a PMS/GMS surgery and therefore whether or not they follow the BMA model contract. Under the model contract, the 2009/10 figures for a basic salary range for a full-time GP (e.g. working 37.5 hours a week or 9 sessions) is £53,249 to £80,354, which equates to between £5,916.55 and £8,928.22 per session. In addition, a salaried GP's pay must also increase each year in line with the DDRB recommendation. In 2009 the DDRB recommended that the minimum and maximum salary range be increased by 1.5% (hence in the year 2008/9 the minimum had been £52,462, 1.5% less than the 2009/10 figure).

GPs who choose to work less than full time (e.g. fewer than 9 sessions) should be paid pro rata, for example five sessions = 5/9 of the full time salary.

Don't forget that you will actually be costing your employer more than just the salary which you see in your wage slip, as they are also responsible for contributing towards your superannuation and national insurance which ends up costing them an extra 25% on top of your salary.

WILL I CONTINUE TO GET AN INCREMENT ON MY SALARY EACH YEAR?

Oh, the yearly increment which we all know and love. The GPC recommends that all salaried GPs receive an annual pay uplift at least in line with inflation and if appropriate in line with the recommendation of the DDRB. You may also want to request that you receive an additional pay increase to reflect your increased experience each year, although this would be at the discretion of your employing practice and should be written in to your contract.

WILL I STILL BE A MEMBER OF THE NHS SUPERANNUATION SCHEME AS A SALARIED GP?

Yes, as long as your employing practice is recognized by the Pensions Agency as an NHS employer.

HOW MANY HOURS WILL I WORK?

This depends on how many 'sessions' you are contracted to do. Under the model contract, full time is defined as 37.5 hours per week which is made up of nine sessions, each of 4 hours and 10 minutes in length. Your exact hours, however, can be negotiated between you and your practice but must still be carefully defined in your job plan. Further guidance on what should be included in this job plan is available on the BMA website but in brief it should outline the employee's normal duties, workload, and important non-clinical roles undertaken within work time, e.g. practice meetings. Any GP who works over and above the 37.5 hours must also have their pay adjusted pro rata or have this recognized by time off in lieu.

ARE SALARIED GPS AFFECTED BY THE EUROPEAN WORKING TIME DIRECTIVE?

In short yes. The EWTD states that a worker should only be required to work a maximum of 48 hours per week on average, although they can choose to work more if they sign a waiver agreement. They also have the right to a minimum of a 20-minute rest break when the working day is longer than 6 hours.

HOW MUCH ANNUAL LEAVE WILL I GET?

Under the model contract, salaried GPs are entitled to at least 30 working days' annual leave per year, plus 10 statutory and public holidays per year (which includes two NHS days that can be taken at any time). Obviously for part-time staff these annual leave entitlements should be calculated pro rata for the amount of sessions worked. It is also good practice to provide additional time off for part-time salaried GPs whose normal days of work do not fall on a bank/public holiday.

WILL I BE ENTITLED TO MATERNITY LEAVE?

Yes. All employees are now entitled to 26 weeks of ordinary leave plus 26 weeks of additional leave regardless of how long they have worked for their employer. Whether you will be entitled to MA (Maternity Allowance) or SMP (Statutory Maternity Pay) will depend on the timings of when you got pregnant and how long you had been working for your employer at that time. Both are payable for up to 39 weeks if the qualifying criteria are met.

Under the model contract, salaried GPs working either full time or part time should be entitled to paid and unpaid leave of 52 weeks in total if they have 12 months or more of continuous NHS service at the start of the 11th week before the expected week of childbirth. Sounds complicated doesn't it? Here's the science bit.

EXAMPLE:

If your baby is due on 10/11/10, the 11th week before your EDD runs from 30/08/10 to 05/09/10 and you will therefore have needed to provide continuous NHS service in the 12 months preceding this date (e.g. from at least the 30th August 2009).

HOW MUCH MATERNITY PAY SHOULD I BE ENTITLED TO?

If you fulfil the criteria in terms of the 12 months NHS service rule, then you should be entitled to the following under the salaried GP model contract:

- 8 weeks of full pay minus SMP/MA
- Followed by 14 weeks of half pay plus SMP/MA (as long as the total doesn't exceed full pay)
- Followed by 17 weeks of SMP or MA (e.g. 39 weeks in total)
- The employee will then be entitled to take the remainder of the 52 weeks (e.g. 13 weeks) as unpaid maternity leave

All of this however, relies upon you filling in the appropriate forms and informing the right people before the end of the 15th week before the expected date of childbirth (e.g. 02/08/10–08/09/10 for the above example) otherwise you may not get what you are entitled to. By this I mean making sure that you get your MAT B1 signed by your midwife and off to your employer along with a letter stating your intended start date for maternity leave by the end of your 26th week of pregnancy.

Since the model contract was introduced however, maternity provisions for hospital doctors have improved to that of 18 weeks of half pay plus SMP/MA after the initial 8 weeks of full pay, so it may be that individual salaried GPs and employers may wish to consider renegotiating their individual contracts in order to reflect this change (although it's unlikely that a practice will want to pay you more than they have to!).

WHAT COUNTS AS NHS CONTINUOUS SERVICE?

The legal view is that any work as a GP principal, salaried GP, or locum should count towards NHS service in the same way that working for a PCO or NHS hospital does. Problems only occur if you have had a break in your NHS service of more than three months without a reasonable explanation such as maternity/paternity/adoption leave or employment under the terms of an honorary contract. Working abroad as part of postgraduate training or for a recognized voluntary services organization is also allowed but for a maximum of 12 months. If the break in service was not due to any of the above reasons then on returning to work the clock must start again for accruing NHS continuous service.

WHAT'S THE DIFFERENCE BETWEEN SMP AND MA?

In order to qualify for SMP you must have been working for your current employer for at least 26 weeks when you enter in to the 15th week before your expected week of childbirth. If you do not meet these criteria then instead you should be entitled to MA. The monetary value of the two is identical, but you will pay tax and national insurance on SMP and you may not on MA depending on your circumstances. MA is also what you will be entitled to should you have been earning more than the minimum amount set by the DWP on a self-employed basis.

WHAT ABOUT SICK LEAVE?

All employees are entitled to SSP (Statutory Sick Pay) from the 4th day of any period of sickness up until a maximum of 28 weeks (and only for days that you would normally have been working). It cannot be claimed at the same time as any other type of benefit such as SMP, SPP, or adoption pay. Under the model contract, salaried GPs are offered the same sick leave allowances as hospital doctors which are as follows:

- First year of NHS service = one month's full pay (after completing four months' service) + two months' half pay
- Second year of NHS service = two months' full pay + two months' half pay
- Third year of NHS service = four months' full pay + four months' half pay
- Fourth/fifth year of NHS service = five months' full pay + five months' half pay
- More than five years NHS service = six months' full pay + six months' half pay

The years of service for sick pay is calculated by aggregating all previous periods of continuous NHS service that have occurred without a break of more than 12 months.

HOW MUCH STUDY LEAVE WILL I GET?

Study leave in the grown-up world of qualified GPs is known as continued professional development (CPD) time and as a full-time salaried GP working under the model contract you are entitled to at least four hours per week on an annualized basis of protected time. This time is to be used to address those specific educational needs that you have detailed in your NHS appraisal and PDP but may also be relevant to the priorities of the practice. How the time is taken needs to be negotiated between you and your practice but can be taken in the form of hours, days, or blocks of days depending on what is required to meet your educational needs. Salaried GPs working part time will be entitled to the same amount of CPD time but on a pro rata basis. CPD activities may include anything from analysing and auditing practice QOF data to attending an update on family planning.

WILL I NEED TO DO AN APPRAISAL?

It is compulsory for all NHS GPs to participate in NHS appraisal. For salaried GPs, time must be set aside within working hours for this to be completed, and this must be in addition to the allocated CPD time. The appraisal interview itself should take place outside of the CPD time but conducted during normal working hours (if it is decided that it will be conducted outside of normal working hours then this is at the discretion of the salaried GP and appraiser, and the salaried GP is entitled to appropriate financial reimbursement or time off in lieu). Appraisals for salaried GPs should be funded by the practice as this is included within their global sum.

WILL I HAVE TO PAY FOR MY LMC MEMBERSHIP?

No. Under the model contract, this should be paid for by your practice or PCO.

WILL I BE EXPECTED TO WORK EXTENDED HOURS?

This depends on what your practice offers and what you agree to on accepting the salaried post. As stated in a previous question, under the terms of the model contract, your employer is required to either pay you extra for these additional hours pro rata or credit with you with time off in lieu.

WHAT IF THE JOB BEING OFFERED IS WITH A PMS OR APMS PRACTICE? WHERE DO I STAND?

If the practice that you are planning to join is not GMS then just be very careful. Neither PMS or APMS practices are obliged to adhere to the model contract and therefore there are no guidelines or rules that they must follow (although when the model contract was produced it was hoped that these practices would try to get their contracts as close to this one as possible).

WHAT ABOUT THE POLYCLINICS?

Polyclinics are large health centres run either by private companies or groups of local GPs in an attempt to provide a general practice service in areas that are considered to be short of GPs (e.g. inner cities).

They tend to be open for longer periods of time each day than your average GP surgery and therefore the jobs that are offered there are based on a shiftwork pattern. They vary enormously in what they offer and what they pay and this has caused considerable debate over the last couple of years. Whilst some are offering NHS-type contracts with good pensions plus maternity and sick pay, others are offering none of these things and the pay doesn't appear to be great despite the antisocial hours and shiftwork that is expected of you.

For some, however, it may be a tempting option as the shiftwork may mean that child care arrangements are easier and it gives you the opportunity to work evenings and weekends if that's something that would appeal to you. For most though, it seems to defeat the point of going in to general practice in the first place as you could argue that the lack of continuity of care and the unsociable hours involved makes it seem more like working in an A&E department or an out of hours centre.

However, as more of these centres get built across the country and more jobs become available in these centres rather than in the form of partnerships or salaried jobs at existing practices, it's likely that a significant proportion of qualifying GP trainees will end up spending at least some of their career working in one of these polyclinics. As with everything there will be pros and cons of doing this and it's perhaps too early at this stage to assess whether they will be a good or bad thing for general practice as most of them have only been up and running for about a couple of years. Watch this space on this one and keep your eyes peeled for more info on things to come—it's a pretty hot topic in all of the GP publications and has been for some time and no doubt there will be further debates over the coming months. If you do, however, see an advert for a polyclinic, don't discard it straight away. Some offer good packages and incentives and will suit some doctors—it's up to you to decide whether you're one of them.

WHERE CAN I GO FOR MORE INFORMATION ABOUT BEING A SALARIED GP?

The BMA has produced a booklet called the *Salaried GPs' Handbook 2010* which is available free to all BMA members and provides detailed guidance for salaried GPs and GP employers. However, if you have individual questions that need answering then further expert employment advice can be obtained by contacting the membership team at support@bma.org.uk or telephone 0300 123 1233. This, however, only applies to BMA members.

> Tip: Before signing any contracts for your new post, make sure that you have checked whether you are happy with the job plan, maternity/sick/annual leave entitlement, NHS pension arrangements, and CPD entitlement. If you're not sure and you're a BMA member, take advantage of their free contract checking service, as once it's signed there's no going back!

'VTS training gives a really good foundation for your future career, but you are not the finished article when you receive your CCT. Very few people walk into their ideal practice. The important thing when choosing either a partnership or a salaried post is whether it gives you room to develop both your own style of practice and within the surgery as a whole. It is challenging, exciting, terrifying, exhausting, but should hopefully work out in the end. I think (hope) that after the first five years things should settle down and I can be well established.'

Newly qualified ST3

Locum work

WHAT IS A GP LOCUM?

A GP locum is one that works on a self-employed basis for various different practices, as and when they are required. Locum work can be anything from a one-off session to 12 months of eight sessions per week, depending on what cover the surgery require. You are paid directly from the practice for the work that you do and do not get any additional benefits such as indemnity cover or GMC contributions. It is your responsibility to find the work and this can be done either by word of mouth, your own marketing, or via a locum agency. Remember that a locum agency will take a cut of any money that you earn so you won't earn as much doing it this way. Work done via locum agencies also isn't NHS pensionable meaning that you also lose out on this front. For more information on locum work see **CHAPTER 21, 'SURVIVING AS A LOCUM (AND MAKING SURE YOU FILL IN THE RIGHT FORMS!)'**.

> 'Doing a longer-term locum post such as maternity cover has its pros and cons. The pros are:
>
> - Security—you tend to be there between three and nine months
> - Being part of the team—getting to know your colleagues and get included in practice 'outings'
> - Administration—time to become familiar with referral processes/forms etc.
> - Follow-up—can deal with longer-term problems, knowing that you will still be there next week/next month etc.
>
> But there are also the cons:
>
> - Tie-in—you are restricted from applying for other positions that become available in the interim
> - Rota—you are usually included in the full rota, including 'Duty Doctor' sessions. This may (or may not) be more stressful than 'appointments only' sessions
> - Administration—compared to short-term locum sessions you will have a larger component of dealing with results/letters/messages etc.
> - Heart sinks—there's plenty of time to accumulate a following!
>
> *GP Locum (1st year post qualification)*

WHAT IS A GP RETAINER?

The Doctors Retainer Scheme was introduced in 1969 as a way of enabling doctors who were taking a career break to stay in general practice on a part-time sessional basis. It initially only allowed doctors to do two sessions per week but was revised in 1998 to allow approved practices to employ a GP on the retainer scheme for up to four sessions a week. All of these four sessions could then be partially reimbursed by the local PCT to the practice. GP retainer scheme posts can be very strict in their criteria and if you are interested in one of these then it is worth checking out your local deanery website to find out further information.

Other types of work

Of course there is more out there than just clinical work in general practice. Whether you want to do something in addition to general practice or if you've had a change of heart and decided that you'd like to follow a different path then there are lots of alternative things to consider.

Teaching

If you've always wanted to get more formally involved in teaching then now's your chance. Being not too long out of medical school and having just spent the last three years sitting exams and completing assessments, you are not only perfect teacher material but you can relate to what the students are going through whilst feeling satisfied that you have finished it all!

There are many different types of teaching posts available depending on whether you have a medical school nearby or whether you have a locally run VTS course. Both of these are excellent ways of getting yourself in to a teaching role (although be prepared for the drop in salary when comparing to clinical GP work!).

Medical education posts

Many medical schools offer posts for qualified GPs to work within the university teaching various aspects of the curriculum to undergraduates in both the pre-clinical and clinical years. This work is either based at your local medical school or your local teaching hospital and varies from teaching physiology and anatomy to teaching communication skills. The pay for this sort of work per session isn't great when compared with the sessional rate for clinical GP work, but it provides an exciting alternative to seeing patients and will add significant credentials to your CV.

GP educator

GP educators are more fondly known as your VTS course organizers or Training Programme Directors (TPDs) and are responsible for making sure that you receive adequate training and support throughout your GP training years. They normally come from all sorts of backgrounds and are of differing ages and skill mix but share one common stem—a love of trying to organize you lot!

Opportunities to become GP educators don't crop up that often but with the recent drive to change VTS study time to involve core groups rather than large group teaching then there are likely to be more openings for course organizers to join the team.

The number of trainees in general practice seems to be increasing year on year and therefore as each cohort grows so must the teaching staff.

Unlike academic teaching, the pay for working as a GP educator is similar to that of working as a salaried GP, but just a word of warning: it's not one of those jobs that you can do from nine to five. Whilst it may seem like an easy ride, it involves lots of organizing, meetings, and planning and may take up considerably more of your time than the one to two sessions that you are being paid for. The recommended salary for 2009/10 is £80,195 which pro rata per session works out about £8,019.50.

Hospital medicine

Just because you've qualified as a GP, it doesn't mean that you have to rush in and work as a GP straight away. There are still a few opportunities for staff grade posts in various hospital specialities if you feel that there are areas of medicine in which you would like to get more experience or polish up as a special interest. The pay won't be as good as what you would get as a GP but it could be a good experience if you fancy a break from general practice for a while and a chance to do something a bit different.

Clinical assistantships

If you're keen to work as a salaried GP but would also like to pursue an additional interest with a view to perhaps becoming a GPwSI, then a clinical assistantship may be worth thinking about. There are plenty of opportunities to do this in various different specialities such as palliative care, gynaecology, acute medicine, and dermatology. In fact there isn't really any speciality in which you couldn't consider becoming a clinical assistant although it depends largely on what service provision is required in your local area and whether or not you can get it organized. Bridging the gap between general practice and secondary care is believed to be the way of the future and clinical assistantships and GPwSI posts may go some way towards helping this. If you have a speciality in mind then start thinking about it early on in your training and investigate options of how you might be able to pursue it. Advice on how to organize this can be found in Box 19.1, which was written by a newly qualified GP who managed to arrange herself a clinical assistantship in dermatology at her local hospital.

Box 19.1 : How to organize a hospital clinical assistant/ GPwSI post

If you have a particular area of interest relating to a hospital speciality, start expressing your interest early. You can start by requesting to attend clinics during 'self-study' time in GP placements. This helps you to improve your knowledge but also enables you to get to know the hospital team. You will need to be

➲

under an honorary contract with the trust (which can be arranged with Human Resources) and should also inform your medical defence provider, although this will not increase your premiums.

Once you are a familiar face in your chosen department, ask if they need any additional clinic sessions. A lot of hospital departments are struggling to meet the demands of the 18-week government target and will welcome your input. Theoretically, you could do this during your VTS, but most trainers would advise against taking on any 'outside' work which might interfere with your training or assessment preparation.

Extra sessions are generally paid on an 'ad hoc' basis, even though you may be doing them regularly. The old 'clinical assistant' pay scale still exists but for most of those exiting the VTS, you will find it woefully inadequate in comparison to what you can earn doing GP sessions. Some hospitals will pay you as a staff grade doctor, which carries a more workable pay scale. Don't be deluded though; you will still be getting about half the amount per session than the GP equivalent. If you are looking for an additional interest to generate lots of income, this isn't it!

Dr Becca Hadzikadunic, GP Locum and Clinical Assistant in Dermatology

Working abroad

If you fancy going to work abroad for a few months then check out the back of the *BMJ* careers magazine. Here you will find advertisements for GPs in countries such as New Zealand or Australia which may be an attractive option if you have no major commitments in the UK. Remember, however, that if you plan to work abroad for more than 12 months then it is likely to affect your NHS pension and continuous service in terms of leave entitlement. You will also need to check whether or not your medical qualifications and professional indemnity are valid within that country without having to do any additional training or pay any additional registration fees.

Voluntary work

After working hard for the last three years to get your MRCGP postgraduate qualification, you may feel that you would like a break from UK medicine and instead might fancy spending some time volunteering for the various organizations across the world that offer help and support to those countries that need it most. This may not necessarily be in the form of medical work as there are many different volunteering opportunities available, depending on where you want to go and what you would like to do. For more information check out the Medicins Sans Frontiers website at http://www.msf.org.uk.

How do I get a job?

When should I start applying for jobs?

Start **looking** as early as possible. Whilst you might not want to **apply** too early (as only a few practices will be able to wait for months until you qualify), it's worth getting a feel for the type of jobs out there so that you can start to polish up your CV to reflect the information that practices will be looking for in your application.

If, however, your dream job crops up and your CCT date is still a way off, you don't really have anything to lose by contacting the practice manager and showing an expression of interest. Take the opportunity to go and have a look round (if they'll allow this—some practices only let shortlisted applicants have an informal visit) and try to judge how soon they need to fill the vacancy and whether or not they would be prepared to wait for the right applicant. Some practices will be brutally honest and say that they need someone quickly and therefore could not wait, whereas others may not have a specific start date in mind and would consider waiting for you if you came across as the ideal person for the job.

In general terms though, you should really be starting to look around six months before your CCT date. This gives you time to look around a few different practices in order to work out what you want and saves you from accepting a job that isn't ideal because you've left it until the last minute.

Where should I look?

Probably the best place to look is either the paper edition of the *BMJ* (although make sure that you are definitely getting the general practice version) or the online version at http://www.bmjcareers.com. This allows you to search for the type of job that you want in the location that you want, but try not to be too picky on the other criteria that you are offered. Chose 'GP and PCT jobs', pick a region and then leave it at that otherwise you will get very few results. Unfortunately you will notice that many of the adverts are for generic locum agencies that operate throughout the country but don't lose hope, in amongst that lot there should be a variety of general practice jobs within your area.

It's also worth registering with the NHS jobs website at http://www.jobs.nhs.uk, who will then email you when an appropriate job becomes advertised on their site. Unfortunately, despite giving your specific criteria when you register, it's all computer generated so you may be bombarded with lots of inappropriate job adverts that don't even have GP in the title. One week I had three emails about jobs as a health care assistant, IT support worker, and senior cardiology nurse which was slightly frustrating but not the end of the world.

If your VTS scheme has a website or social networking group then don't forget to look there for potential job opportunities. Many surgeries chose not to advertise nationally because they would rather have a local candidate and therefore might send out emails via your TPDs or post vacancies on the website.

How do I write a CV?

Given that most of your jobs to date will have been organized via the Foundation years allocation programme and the general practice recruitment system, some of you may never have written a CV. Those who have had additional jobs outside of medicine or who have worked in other specialities before starting general practice may well have written one, but it is likely to need considerable tweaking before using it to apply for any general practice jobs. In the world of real general practice, traditional CVs are very much still alive and kicking with most job adverts requesting that you send a cover letter along with a copy of your up-to-date CV to the practice manager. It is therefore important that you devote a sufficient amount of time to making your CV look good, as it will be the first thing to make an impression when you apply for a job (that is, if you haven't already visited the practice and wooed them with your wit and charm). Make sure that on first glance your CV looks smart, is easy to read, and contains all of your relevant personal details. This means printing it out on good quality paper, avoiding the use of fancy illegible fonts, aligning the text, doing a spell check, and making sure that it is free of tea cup stains before sending it. The personal details section should include name, postal address, email address, mobile or landline number, DOB, nationality, GMC number, medical defence organization membership number, Hep B status, Performers List number, and date of last CRB check. It is also worth writing a few sentences below this information to give the reader a brief outline of what your aspirations are and what you like doing in your spare time, as a way of giving your CV a bit of a 'personality'.

Next should come a list of your **relevant** qualifications with the most recent first, e.g. MRCGP and the dates of when they were achieved. You may or may not want to also include in this section your A-level and GCSE results although they probably aren't really relevant this far down the line!

You will then need to list your employment history, again starting with the most recent job first. You may choose to include in this a brief summary of what each job involved although remember that you need to keep it brief so this may not be possible. This section should include the title of the post, where you were based, and the dates over which you occupied that role.

Following this you should record any additional courses you have attended, certificates that you may have achieved, additional skills or positions of responsibility that you may have, along with any evidence regarding special interests which you may be offering.

If you have any publications, research, or audits which are particularly relevant then you may also want to include a brief section explaining these, although remember that unless relevant they may just make it look as though you are trying to pad out your CV.

Finally you should name at least two referees (one of whom ideally should be your current boss) and provide accurate contact details for them. Make sure that you choose referees that will be happy to 'sell' you to a prospective employer. Whilst this may seem obvious there are several people who will chose a reference based on how impressive the name looks on the page rather than because of the positive things that they can say about them.

Once completed, make sure that you spell check it, read it through carefully to ensure that you haven't made any glaring errors, and print it on decent quality paper (unless they have requested to receive it by email!). It's also worth asking your educational supervisor or trainer to read through it so that they can give you an opinion from a potential employer's point of view.

Box 19.2: Top tips when writing your CV

- Tailor it to the job
- Make it look smart
- Cut out the junk
- Keep it simple
- Don't lie
- Avoid fancy fonts
- Keep it to a maximum of three A4 pages
- Use good quality paper
- Chose reliable referees
- Don't forget the cover letter!
- And check it before sending it!

What should I look for in a job advert?

You can gain a great deal of information about a practice from their job advert so it's worth having a look at a few to see how they compare. Some will keep it very brief, only detailing how many sessions they require, the list size, and when they would need the successful candidate to start. Others will give a short summary of the practice and give suggestions about the general ethos of the place by using words such as 'friendly' and 'busy'. Read into these what you will but essentially you will only really get to find out what the place is like by going to visit. Some practices openly accept informal visits whereas others will actively refuse this, allowing only those who get an interview to step foot inside the door. This approach can be off putting for prospective candidates and may give the impression that the practice has something to hide. In truth however, it is more likely that they may have been inundated with applicants and therefore simply don't have the time to show lots of people around the practice in the weeks running up to the closing date.

When looking at the advert, think about what it is you are really looking for in a practice. If you have your heart set on working part time in a small practice out in the country, then you probably won't want to apply for a full-time salaried post in a 12,000-patient inner city practice. But that doesn't mean that you should definitely exclude it and if you are able to go and have a look around then do.

In general, a great deal can be gained about what the practice are looking for by the wording of the advert, e.g. 'special interest in gynaecology and family planning ideal' may suggest that they are looking for a female or '10-session partner needed for busy inner city surgery with extended hours' may well mean they are looking for a male. Of course these are sweeping generalizations and there are plenty of women who would happily work 10 sessions in a busy inner city surgery, but in these days of sexual equality when the practice isn't allowed to specify the desired sex of the applicant, they will often give clues in the advert as to what they are really looking for. And of course this is understandable—most surgeries will want to offer their patients a choice of which doctor they can see and therefore it's only natural that a practice would want to have a variety of doctors in terms of age, sex, and background.

Box 19.3: Questions to think about when looking at job adverts

- How many sessions?
- Size of practice?
- PMS/GMS?
- How many partners?
- Rural or town?
- QOF achievement?
- Start date?

Very few job adverts will mention salary although some will state that they will pay in accordance with the BMA contract if it is a salaried post. Remember that PMS practices are not bound by the BMA contract so can offer you whatever they like in terms of salary and leave commitment.

Where can I find out more information about the practice?

Most practices will have their own website where you can find out basic information such as what time their appointments run from and until, how many partners there are, and what additional services they offer, such as physiotherapy, antenatal care, and counselling. If it's a training practice then you will be able to gain a vast amount of knowledge from their current and previous ST doctors (registrars) although you may want to take some of what they say with a pinch of salt, as it is likely to be very different working there as a fully qualified doctor within the practice when compared with being the trainee.

Probably the best way to find out more information is to approach the practice manager, even if it's just by telephone, as not only should they be able to give you more information about the practice but they will also be able to give you more specific details of what the job entails and what kind of person they are looking for.

Who should I approach first at the practice?

This depends on what it says on the job advert. Normally your first port of call would be the practice manager although sometimes the senior partner will be named as the point of contact. Remember that the impression of you starts from that first phone call to the practice, so whether it is the receptionist that you speak to or the senior partner, make sure that you sound professional, are polite, and know exactly what you want to say. I have heard a few stories of applicants' CVs being put straight in the bin as soon as they arrive, following the applicant either being rude to the receptionist when calling to make enquiries or sounding completely clueless when on the phone to the practice manager.

Informal visits—what are they?

Informal visits don't do exactly what they say on the tin. Whilst they are informal in the sense that you will probably be able to go at your own convenience and won't be subjected to a grilling by a panel of doctors, they may well be sussing you out in an 'informal way' from the minute you step through the door. For this reason, it's probably best not to turn up late or in your jeans and t-shirt, as the practice manager is likely to feed back to the partners what their first impressions of you were. Make sure that you are polite to the reception staff when you make that first phone call and at the visit appear enthusiastic, professional, and reliable, even if you decide at first glance that it isn't quite the right practice for you. Informal visits can last anywhere from a few minutes to an hour or more, depending on who shows you around, what questions you may have, and how interested you are in them and vice versa. You may be introduced to one or more of the partners and may feel that you are undergoing an informal 'interview' rather than an informal 'visit'. Try not to let this faze you. General practice isn't so much about getting the most qualified person for the job, it's about getting someone who will work well within the current team and therefore personality counts for a lot. Remember also that as with any job, it has to be right for both the employer and the employee so don't be frightened to ask questions and have a good look around.

Tip: Make sure that you are polite to the reception staff when you make that first phone call and at the visit appear enthusiastic, professional, and reliable, even if you decide at first glance that it isn't quite the right practice for you.

What does 'with a view to partnership' actually mean?

This usually means that whilst the practice are ideally looking for a new partner to join the team, they wouldn't want to enter into such a big commitment before knowing whether the successful applicant is going to 'fit in' at the practice. They therefore may offer the job as a salaried post to begin with, but then if all works out well, the intention would be to make that person a partner. However, be very careful of this one. Whilst most practices are genuine in making this statement, there have been circumstances where doctors have been offered the situation, only to find two years down the line that there was no intention of them ever being made a partner. Make sure therefore that you ask about this at your interview. Ideally they should be able to give you some guidance on when they would hope for you to become a partner, but if they're vague and don't really have answer for this then be suspicious. Some practices may use this phrase to entice people to the job and actually have no intention of making the successful applicant a partner. You could argue, however, that the trial period is as much for you as for them as in a similar way, you probably wouldn't want to enter straight into a partnership (which is considered a bigger commitment than marriage!) without having some idea of what you are letting yourself in for.

When will I get partnership status (and money!)

This depends on what you have agreed in your contract but it is likely to be at the very least three to six months before you are made a fully-fledged partner. The money side of things will depend on whether you have bought in to the practice or whether you will just be receiving a proportion of the profits and again will vary hugely from place to place. The most important thing is to make sure that you know where you stand and what is expected of you from the very start.

I've sent out my CV for lots of jobs but haven't been offered any interviews—help!

Hmm… perhaps spend an hour or two with your CV and look at whether it is really selling you or not. You should also ask your educational supervisor or even just a friend to look over it and give you an honest opinion of where you may be going wrong. Also think about whether it relates to the jobs that you are applying for. Sending out your CV for every job that crops up is likely to make you look desperate and probably means that you aren't spending enough time tailoring your CV and cover letter to the job description. Generic cover letters and CVs are very easy to spot and therefore even if you are applying for three jobs a week, you mustn't let it come across that this is the case. Make sure that your cover letter is addressed to the right person and not to the practice manager of the surgery down the road that is also advertising a post. This will not only make you look disorganized but it is likely to mean that your CV will go straight in the bin. Also think about whether you have the necessary skills and attributes required for the job. Some adverts will read, 'the successful applicant should have…' or they may even have written a person specification document. Make sure that you read these in detail before applying. If they are looking for someone who has at least three years of post VTS experience with a special interest in minor surgery, then they will probably not be interested in someone straight out of VTS who has never even done a minor surgery course. Of course, in terms of special interest there is always scope with this, as if they like you and you meet all of the other criteria then this may be something that the practice can train you up in to their advantage. Also don't be put off by the fact that you are soon out of GP training. This can be very appealing to a surgery who are keen for some 'fresh blood' and would love to have an enthusiastic new doctor who is up to date on recent changes in medicine and who hasn't been doing it long enough to be cynical! (They can also pay you less than more experienced GPs which may be an additional advantage!)

I've still got a few months of VTS to do—will they wait for me?

Very possibly. It depends how desperate they are to fill the post and how keen they are to take you on board. Some practices won't be able to wait for the right candidate (e.g. PMS practices) as they may lose the funding for the post if they don't fill it within a certain time period. Other practices, however, would much rather make sure that they get the right person for the job and if this means employing a locum for a few months whilst you complete your VTS then so be it. Make yourself worth waiting for and you'll be fine!

The interview

What should I wear to the interview?

This is a very personal decision but ideally you should look smart and feel comfortable. Your interview may last anything from 20 minutes to over an hour so squeezing yourself into a suit that you bought for your medical school interviews probably isn't the best idea. Think about the first impression you will make when you walk in the room and ask someone else's advice before turning up in your favourite crushed velvet blue jumpsuit.

What should I take with me to the interview?

You may be asked to bring specific documents with you to your interview and if so these should be detailed on your interview letter. If nothing is mentioned then at the very least you should take your GMC certificate, your MRCGP certificate, proof of your medical indemnity, proof of Performers List status, proof of your Hep B (and possibly HIV) immunity, driving licence, and car insurance certificate. You should also take any certificates which are relevant to the qualifications stated on your CV such as DRCOG or DFSRH. They probably won't ask to see them but it looks professional if you have it all to hand in a smart portfolio.

How do partnership interviews differ from salaried GP interviews?

They are likely to be longer and may require you to do significantly more preparation in advance. Some practices ask prospective partners to give a presentation as a way of helping the candidate feel relaxed by coming to the interview with something which they have prepared. The topics can range from something personal to what you could do to improve the practice. Other practices may have more inventive ways of making their choice by inviting you out for dinner or taking you to a show! Recruiting a partner is a much bigger decision than recruiting a salaried GP and therefore you will probably be worked harder during this interview. The decision is also likely to take longer when recruiting a partner and may involve rounds of second or even third interviews in order for the practice to make sure that they are choosing the right candidate.

Should I ask to see the practice accounts and when would this be appropriate?

If you are applying for a job that may lead to partnership then yes you probably should ask to see the accounts. However, it wouldn't be appropriate to do this on your informal visit, or before you have applied for the post. In fact, it's probably only something which is appropriate once you have got through to a second interview and should be broached sensitively and professionally. If you are planning to do it however, then you should know what you are looking for and therefore asking the advice of an accountant is probably essential at this stage.

What will they ask at the interview?

They will probably want to know a little more about you and why you specifically want this job. This is where you may fall down if you have applied for every job in the locality for the last three months! Even if this is the case, then make sure that you prepare your answers for simple questions such as these as they should be easy points

Box 19.4: Examples of possible questions which may be asked at interview

- Describe your career to date
- Give of an example of when you have worked well within a team
- Tell me about one of your heart sink patients
- What skills can you offer to our practice?
- What does CPD mean to you?
- What are your weak points?
- Where do you see yourself in 10 years' time?
- Is there anything about you that we need to know but that you haven't told us yet?
- What makes you tick?
- What attracted you to this job?
- Describe one of your heart sink patients and how they make you feel

to score. They may ask you about where you see general practice going in the future or what additional things you could offer to the practice. Make sure you think about topics that are relevant in the press and within GP world as they may be interested to hear your opinions on some of these. Essentially they will be deciding whether you are the kind of person that they would like to have within their team, so try to relax and allow them to see some of your personality.

Will they be testing my medical knowledge at the interview?

Probably not but they may well ask how you would deal with a few difficult case scenarios (e.g. ones that have ethical, moral, or confidentiality dilemmas). They are likely to be questions that don't have right or wrong answers, so there's no need to memorize the *Oxford Handbook of General Practice* the night before the interview.

I'm pregnant—should I mention it at the interview?

Hmm... a difficult one. This depends on how pregnant you are. If you have a pear-shaped mass protruding from your belly then you may not have to say anything because it will be obvious. In this circumstance you must definitely mention it as they will appreciate your honesty and can talk openly with you about how it could work if you were to be successful in getting the job. And it might not necessarily affect your chances. Women have babies, it's a fact of life, and if you're the candidate that they've been looking for then they may be willing to wait until after you've had the baby, or even employ you before then.

If you are not physically obviously pregnant and you are less than 26 weeks then you are not obliged to tell anyone. However, whilst legally you do not have to say anything, if you accept the post without saying anything and then a few weeks later send them your MAT B1 form stating when you intend to go on maternity leave, then you may find yourself in a slightly awkward working environment. It's probably better to be honest from day one to prove to your new employer that you are trustworthy and reliable. If you are only in the early days of your pregnancy and have not yet had your 12-week scan then it is perfectly reasonable to keep it to yourself as at this stage you may not even have told your family let alone a potential employer. Anything could happen at this stage and whilst officially an employer is not allowed to discriminate against you because you are pregnant, it may well affect your chances of securing the job.

What questions should I ask at interview?

You will probably have had many of your questions already answered during the informal visit or chats with the practice manager but don't be afraid to ask anything that you feel is relevant. It would be perfectly reasonable to ask about rate of pay, expected on-call and extended hours commitment, along with when they would require you to start the job.

How many people will they interview?

This will vary hugely from place to place and will depend on what type of post they are offering. You will probably be able to work it out based on the length of your interview and the number of hours which they are spending doing it. Whilst you could use this information to work out the odds of you getting the job, it's actually not very relevant because it may well turn out that they appoint none of the 20 people that they interviewed as there wasn't anyone right for the job.

How many people will interview me?

Anything from two upwards. It would be unusual for you to be interviewed by a single person and there probably won't be more than five or six as that would be far too intimidating for the candidate. If it's a partnership interview, however, then you may be interviewed by all of the partners plus the practice manager which if it's a big practice could be rather more than that!

My friend has an interview for the same job as me—help!

This is likely to be a fairly regular occurrence and will become the norm as the number of trainees increases disproportionately to the number of available jobs. Try not to let it faze you as things will normally work out for the best. Whilst it may be disappointing and disheartening seeing your friends secure jobs that you would have liked for yourself, remember that something else will come along which may be even better suited to you.

REFERENCE

1 BMA. *Model Salaried Contract: GMS Practice*. Available at: http://www.bma.org.uk/employmentandcontracts/employmentcontracts/salaried_gps/SalariedGPcontractGMS0209.jsp (accessed 12 August 2010).

CHAPTER 20

Preparing to leave the nest...

Congratulations! With the exams over and just the rest of your rotation to complete you should be well on the way to becoming MRCGP positive, awaiting only the delightful certificate that you get from the college to confirm your efforts. Be warned however, this certificate not only displays your fantastic achievement but also the photograph that you submit when you apply for your CCT, so make sure it's a good one and wasn't taken the morning after you stayed up all night trying to get your ePortfolio up to scratch ready for ARCP. Of course I am joking, instead it is likely to show you as fresh faced and enthusiastic, ready to start your career as a fully fledged GP! No more form filling and ePortfolio cramming... oh if only!

But be proud of yourself. The last three years have undoubtedly been extremely busy, stressful, and exhausting (not to mention expensive!), however, it should have all been worth it now that you can work independently as a GP and have the letters MRCGP after your name!

Of course, it's not all been about exams. Passing the three components of the MRCGP must also have been accompanied by the completion of three full years of training posts, so if you've had any time out from your training, whether that be for maternity, paternity, or sickness leave (or more exciting OOP adventures overseas!) then make sure that your expected completion date is the same as that dictated by the deanery. Anyone who has had more than two weeks out of their training for whatever reason may be required to extend their training, so if you think that this applies to you then discuss it with your training programme directors as soon as possible. You don't want to end up losing out on your dream job because you underestimated when your CCT date would be.

The paperwork stuff

How do I prepare for my final panel review?

The ten-point checklist:

1. Is the self-assessment of your competencies complete?
2. Has your ES completed the assessment of your competencies?
3. Do you have a CSR for every post with competencies that are specific to that post signed off?
4. Does your PDP look ahead to the next few months of learning needs?
5. Does your learning log show evidence of reflection?
6. Have your DOPS all been verified by a senior clinician?
7. Are your OOH sessions clearly recorded with evidence of the 72 hours being completed?
8. Is your CPR training up to date and signed off?
9. Does your evidence show a balanced coverage of the curriculum?
10. Have you signed off your final review?

When will I have my final panel review?

The earliest that this can be carried out is eight weeks before the end of your rotation, so it is essential that you have completed all of the necessary WBPAs, learning log entries, DOPS, and OOH sessions well before this date. You will probably have your final ESR around a month before the panel date so your ePortfolio should be looking slick and 100% complete around three months before your rotation ends in order for you to be deemed as 'competent' and complete your training (August finishers, for example, should have had their ESR by the end of May ready for the June panel). If you have been successful in completing all of the required elements of your final ESR then your educational supervisor will make a recommendation to the Deanery that you are eligible for your CCT. This, however, is subject to external moderation in the deanery via the ARCP panel. For more information on ARCP reviews and how to make your ePortfolio look good see **CHAPTER 12, 'STOP AND CHECK! ARE YOU READY FOR YOUR ARCP PANEL REVIEW?'**.

How do I check that I have made it successfully through the final ARCP panel?

This can be done by checking the 'Progress to Certification' area on your ePortfolio. If you click on to your final review document you will find two boxes that could have been ticked. The first states 'achieving progress and competencies at the expected rate' and the second states 'has **completed** all of the required competencies for completion of training'. Make sure that the second box has been ticked otherwise you'll encounter problems when applying for your CCT.

What is the CCT?

This is the certificate of the completion of training and once granted allows the licence holder to work in any capacity unsupervised in UK general practice.

How do I apply for my CCT?

Several months prior to your CCT date you will be sent an email from the RCGP asking you to register with the GMC for the online certification process. This can be found at http://www.gmc-uk.org/certificationonline and will initially require you to enter your personal details, agree to some security statements, and 'create an account'. Following this you will receive an email containing your user login and password in order to access the online certification application.

Once you have received this information you are then required to upload several supporting documents into your account including a JPEG colour photograph of yourself, a scanned copy of some photo identification, and a copy of your CV. You will also need to have to hand your NTN number and date of completion of training. The criteria for the supporting documents in terms of JPEG size and resolution are quite specific so make sure you check this before you spend hours trying to upload them incorrectly. Once you have created your account, the documents can be uploaded at any time so it does not require you to complete the process all in one go and instead allows you to 'save', 'exit', and 'return' at any time. As part of the process you will also need to pay the fee for your CCT which in August 2010 was £805.

Once you have paid the fee and all of the required information has been submitted, you are able to track the progress of your CCT application online by simply logging in to your account. The RCGP recommendation stage cannot be done until the final ST3 review has been signed off by your educational supervisor and verified by the ARCP panel.

If this all goes through successfully (which it should!) then it won't be long before your crisp new certificate lands on your doorstep in the post. As soon as it arrives you will need to take a copy to your local PCT so that they can update your position on the Performer's List.

> 'Pick a passport photo of yourself that you at least half like, because it's going on the certificate, and will remain there forever'
>
> ST3

How much do I have to pay to the RCGP for my CCT?

For August 2010 it was £805 although this is likely to increase along with inflation. You will not be issued with your official CCT certificate until you have paid this amount in full and unfortunately there is no option for monthly direct debit payments payment plans.

> 'Finding out how much money we were required to pay to get our CCT was a big shock! We had no idea that we had to pay, let alone that it would be that much. I had only just recovered from paying out the money for CSA!'
>
> ST3

How long does it take for my CCT to come through?

This can be anything from a week to a month so make sure that as soon as you are prompted to register for the online process that you do so. Remember that the final stages of the process are likely to be happening around July time which is not only a peak season holiday month but also is the time when most ST3s throughout the country will be trying to do the same thing. The CCT certificate is what the PCT require to update your position on the Performers List to that of a qualified GP and if you don't get this to them in time, it may mean that your new job start date is delayed.

Is there a graduation ceremony? If so – where?

Yes, sort of. For our cohort we were invited to attend the national RCGP conference up in Glasgow which would be followed by a mini-graduation type ceremony. Very few of us decided to take them up on this offer as it was a long way to go to be given a certificate that you already have and wasn't until the November (by which time it was old news, but that may just be the cynic in me!). As it tends to be incorporated into the RCGP conferences there are other opportunities within the year at alternative locations so keep your eyes peeled for when this might be happening. The college will also email you with possible dates and venues at various times throughout the 12 months either side of your CCT date.

I'm out of sync due to having a baby—when will I get my panel decision?

You will need to check this with your deanery although it should be made clear on your ePortfolio.

Do I need to have an appraisal before I finish?

You don't necessarily need to have an appraisal before you finish but many trainees do, not only so that you get to have a practice run with your trainer (rather than a stranger at your next one!) but it also saves you from having the panic of trying to get it done towards the end of the year. Performers List regulations state that it is compulsory for all NHS GPs to participate in NHS appraisal so the sooner you get into the habit of doing it the better. For more information on appraisals see **CHAPTER 22, 'LIFE AFTER THE VTS (AND WHERE TO GO FOR SUPPORT AFTER LEAVING YOUR TRAINER!)'**.

How do I upgrade my status on the Performers List?

You **should** already be on the Performers List from your time spent working as a registrar and if your new job is within the same PCT then all you need to do is take your CCT certificate to the registrations office and ask them to update your profile from GP registrar to fully qualified GP. This should only take a matter of days and **should** be a fairly straightforward process, although can end up being a little frantic as it relies on you having received your CCT certificate from the RCGP (which may not be until the very last minute!). As soon as you have your CCT certificate safely in your mitts, I would suggest that you take it in by person to your local PCT to ensure that they definitely have possession of it in order to update you on the list. If you're planning to locum, you only need to be on one Performers List (in fact you are only **allowed** to be on one of them!) in order to practise. The easiest thing to do would be to stay on the one that you were on as a registrar but obviously if you've got a salaried post and are moving away from the area then bear this in mind as you come to the end of your registrar year. The reason being that depending on the PCT and whether or not you need a new CRB check this process can take anything up to two months and you don't want to be a faced with someone from the PCT giving you a rather frustrating 'computer says no' face the day before you're due to start work!

If your new job is based within a different PCT and you are planning to be there long term then you will need to transfer your membership to that new Performers List. This can also take time but should not prevent you from working as you should be covered by your previous Performers List until the transfer has taken place. However, never take these things for granted and make sure that you check out what is required well in advance of the date that you are due to commence your new job.

> 'When I got the phone call from the PCT the Friday before I started work as a locum the following Wednesday saying that they couldn't update my position on the Performers List because I had been removed two years ago I could have cried. Not only did this mean that I had been practising 'off list' so to speak for the last 16 months but it meant that I was left 26 weeks pregnant with the only four potential weeks that I could locum being destroyed. I had joined the Performers List two years prior when I did an innovative post and had presumed that once on it I would stay on it until I completed my training. The letter that I had been given at the time made no suggestion that I would be removed from the list once I went back into hospital medicine and they certainly hadn't written to me to let me know this. The PCT couldn't really give me an answer as to why I had been removed and despite many frantic phone calls, chasing the postman to get the forms that they had sent, not to mention

lots of stress and tears I rushed into the PCT office to be told that it could take between one and two months to get me back onto the list. This was despite only just having had a CRB check for the same PCT for an alternative job the week before. Unbelievable.'

Newly qualified ST3

I've just been to take my CCT certificate into the PCT and have found out that I have never been on the Performers List—help! What can I do?!

This shouldn't happen, however there have been several cases of this happening across PCTs nationally in the past. If you are unfortunate enough to find yourself in this situation then there isn't a great deal that you can do. Much of what happens next will depend on the efficiency of the PCT and the time of year. Getting on the Performers List involves filling in lots of forms and taking in documents to be seen and photocopied. The most important parts of it though are the CRB check and the references check which can take anything from 10 days to two months to be approved.

So if you are faced with the horrifying fact that you are NOT in fact on the Performers List and therefore CANNOT work (OR GET PAID!) in general practice for possibly one to two months, then you need to think carefully about how you are going to pay your mortgage and finance the TVR that you just splashed out on thinking you were going to get rich doing general practice. Well, if you have completed your rotation and are now out in the big wide world on your own I would suggest contacting your local hospital to see if you can get any last minute locum work as an ST in A&E/O&G or one of the other specialities in which you have worked. If you were a hard worker and a reliable doctor during your hospital posts your local hospital where you worked would probably be delighted to have you back rather than an unknown locum that has just landed in your area. It is only general practice that you cannot do without being on the Performers List, not other medical specialities, although if you are planning to work in the hospital for a short time then it would be worth letting your medical defence organization know as this may impact on your subscription.

If I'm not on the Performers List is my medical indemnity invalid?

In short—yes. In order to practise as a GP in this country you **must** be on a PCT Performers List. It doesn't necessarily have to be your local Performers List (although ultimately this should be the case) but you must be on one somewhere. Being on an alternative list elsewhere will tide you over until you sort it out locally but your local PCT are likely to insist that you switch over to their list ASAP.

The medical defence organizations therefore have every right to refuse to support you if it turns out that you made a blunder during the time that you were not on the Performers List; however, if you have been paying your membership at the correct rate for a GPST and have not defaulted on any payments then they would be very unlikely to refuse to support you if a claim against you was made. It is a scary prospect though and it would be prudent to let them know as soon as possible if this was the case.

How do I update my professional indemnity?

Updating your status with your medical defence organization is vital and one of the easy jobs (although if you don't have your fees included as part of your new salary then the amount may come as a bit of a shock!). The amount that you pay all depends on how many sessions you are working, what type of work you are doing, and which medical defence company you are registered with.

If you don't have a regular job as yet and are planning to locum for a while then work out how many sessions a week you are planning to work on average and pay for that rate. If you do more or less than this then they can easily adjust your payments or give you a refund if you have paid too much. My advice would be to overestimate the amount of sessions that you might do as if you do ever need to rely on them for assistance and you have been paying for a one- to two-session rate but actually doing seven to ten sessions, they won't look too kindly on you. It's much better to get a refund than to be faced with a huge bill at the end of the year.

Leaving your training practice

For most of you it is likely to be a sad moment when you finally leave your training practice and prepare to step out in the big wide world alone without your trainer. Don't forget though that most trainers and educational supervisors will be more than happy to support you in those first few months, in terms of looking over CVs, guiding you in the direction of forthcoming jobs, and giving you a reference. Hopefully by this point you should feel fairly confident in carrying out your daily practice without having to knock on your trainer's door for advice, but remember, even very experienced GPs don't know it all and need to ask questions sometimes so don't be afraid to do this if required. Most GPs won't mind you asking them for advice, particularly if they are a training practice and used to this and if you get really stuck then there's always 'GPNotebook'—which unlike your trainer (as long as you've got access to the Internet) will always stay with you! As with every new job there will be teething problems when you first start and you will not be the first person to ask where the blood forms are or where you can find a speculum.

> 'Unless you have been the trainee from hell, most trainers like to keep in touch with their protégées and see how they are getting on. They also feel useful if occasionally they are asked for advice or help beyond the training years!'
>
> *GP trainer, Warwickshire*

What do I need to do in the few weeks before I leave?

Make sure that you tidy out your room (but don't steal any of the equipment!), leaving it clean and tidy for the new registrar and hand over any patients that have ongoing issues or for whom you are awaiting results. It's probably better to hand these over to one of the regular GPs rather than the incoming registrar as they will have enough to deal with trying to find their feet in the first few weeks. It's probably also worth booking a holiday or at least giving yourself some time out between the end of your registrar job and the start of your new qualified GP job. The last three years will have been hard work and it would be nice to start on day one with a fresh face and some enthusiasm. Never underestimate the benefit of a holiday. No one is impressed by someone who works themselves into the ground and remember the cheesy cliché that if you can't look after yourself then you will struggle to look after others (e.g. your patients). At the end of your rotation not only will you need (and deserve) a physical break in order to transform yourself from registrar to qualified GP, but you'll also need some time out to make sure that you've signed and sent off all the necessary forms to the PCT to enable you to start work.

What do I need to do before I can start work?

The end of VTS ten-point checklist:

1. Check your position on the Performers List—you will need to make sure that you are on one and that you are listed as a fully qualified GP and not a trainee
2. Update your professional indemnity for the required amount of sessions and the correct type of work (e.g. salaried/locum/partner)
3. Make sure that you have paid your GMC fee for the year
4. Update your status with the RCGP
5. Start a CPD/appraisal folder
6. Sort out any necessary child-care arrangements
7. Make sure that you know your way around the computer system at your new practice (or that you have a vague understanding of all local systems used if you are planning to work as a locum)
8. Buy yourself a doctor's bag and any necessary equipment if you haven't done so already (and don't steal lots of supplies from your training practice as they won't be very impressed!)
9. Change your *BMJ* to the GP edition if you haven't already done so
10. And finally... frame your certificate! And be very proud of what you have achieved!

CHAPTER 21

Surviving as a locum (and making sure you fill in the right forms!)

As you near the end of your training you will probably start panicking about whether or not you will be able to get work and how much you will get paid. After all, for the last three years (at least) you will have been comfortably cocooned within a rotation which promises you not only regular work but a stable income (and a very reasonable one at that!). Although there perhaps aren't as many partnerships out there as we all would like, there are still plenty of salaried jobs and an abundance of locum work (depending on where you live). Whilst the prospect of locum work may seem daunting, there are many advantages of picking and choosing when and where you want to work (not to mention the money that you can make if you are good at it!). Locums notoriously have a bad name which is probably down to the fact that they are perceived by those in permanent jobs to be the dregs of the profession who can't get a 'proper' job. In fact this isn't normally the case and there are many GPs out there who are working as locums because it suits them and their lifestyle and in fact they are very good at what they do.

How to organize yourself as a locum is a subject which often isn't covered well during your training years but is something which is likely to become particularly relevant as you near the end of your VTS rotation. From my experience the limited sessions which were provided on this subject were often very mercenary, centring wholly around the amount of money that you could squeeze out of your struggling, annual leave-burdened training practices. What you really need, in fact, is some first-hand hints and tips on how to get the work and create yourself a good reputation within your local area. Remember, you will probably not want to be a locum forever and should therefore treat every practice that you work at as a potential employer.

Becoming a locum

What are the advantages of being a locum?

Apart from the obvious one of being able to pick and choose when and where you want to work, there is also the added bonus of getting to learn different computer systems and just seeing how other places operate. During your training you will probably only have experienced two different surgeries, and it's not until you've been employed by a few more that you start to realize which things work within a practice and which don't. This is all good research for the future, when you decide to set up your own all-singing all-dancing surgery and are able to incorporate all of the good ideas that you have stolen from other surgeries.

You can, of course, also make a fair amount of money working as a locum, but remember that the amount of work that you can get is rarely consistent and you will probably find that much of it crops up at the same time (e.g. the summer months). It is, however, incredibly flexible meaning that you could work every day one week and none the other without necessarily losing out financially.

Advantages of being a locum:

- Very flexible—in terms of when/where/how you work
- Fewer heart sink patients!
- Well paid
- Less patient admin
- Less involvement in surgery politics!
- Variety
- Getting to experience lots of different surgeries to help you decide what would work best for you (and your family)

What are the disadvantages of being a locum?

Well, you mean apart from everyone assuming that you got your degree on the Internet and being made to go out and do all of the duff visits? Jokes aside, there are plenty of disadvantages to locum work, some of which I've listed below:

- Not knowing any of the patients
- Unfamiliar computer systems
- Not being able to see patients the next day or follow them up in a week's time if you need to (this can cause both psychological angst and lead to an educational deficit)
- Unreliable income
- No paid sickness/maternity/paternity pay
- No paid annual leave
- Having to work in a strange room and provide your own equipment

- No contribution towards professional subscriptions or professional indemnity
- Lots of personal admin and form filling (you need to be organized)
- Increased clinical risk (this is reflected in the increased indemnity price)
- Often required at short notice
- Difficulties with child care

Of course, this list is very generic and doesn't take in to account what type of locum work you are doing. Long-term locums such as maternity cover, end up essentially being similar to having a salaried job although are unlikely to include paid leave of any sort.

Preparation

What do I need to have done in advance?

Most of this is covered in **CHAPTER 20, 'PREPARING TO LEAVE THE NEST...'**, as the same will be relevant to all trainees preparing to start work as a fully qualified GP, however the most important things to get done before you start work as a locum are:

- Check that your status is updated to that of a fully qualified GP on your local Performers' List (and that you are still on it!).
- Update your status with your medical defence organization in order to indemnify you for locum work. This is often more expensive than if you were doing work as a salaried GP as you are perceived to be a higher-risk practitioner. Your indemnity fee will depend on how many sessions you are working on average over the forthcoming year. If you are not sure how many sessions you are planning to work, it's probably worth paying for the five to six-session tier and then upgrade indemnity according to the amount of work that you get. This can be done fairly simply by contacting the membership department by phone who will issue you with a new certificate.

How do I find the work?

There are many ways in which you can do this. The easiest (but not necessarily the best) way is to register with one of the many locum agencies that you will find advertised on the *BMJ* careers website. The problem with agencies, however, is that they take a cut of any money that you earn and you may also find yourself commuting to practices a fair distance from your home and in areas where you don't know the system. Work gained via locum agencies is also unlikely to be NHS pensionable which will have a detrimental effect on your pension long term.

Hunting down your own work is a much more cost-effective option, despite the leg work and phone calls involved. Your TPDs should be able to put you in touch with the local non-principals group which can then include you in any emails relating to locum work required in the area. The TPDs are also the first ones to know about any partnerships or salaried posts on the horizon so it's worth making sure that you remain within their radar and are creating a good impression. Make sure also that you regularly check your VTS website or social networking site as practice managers will often post locum advertisements here via the TPDs.

Another option would be to email or post your CV including a covering letter to all of the practice managers within the area that you would like to work. This can be time-consuming and costly in terms of postage but if your CV looks good and is accompanied by a friendly letter explaining your availability and skills then it should hopefully yield some work. It's likely to be even more fruitful if you're prepared to actually go into practices and deliver your CV by hand so that the practice manager can put a face to the name.

Probably the most common way of finding work, however, is by word of mouth. Keeping your ear to the ground and asking your practice manager, trainer, and salaried colleagues if there is anything coming up is likely to be the easiest and quickest way to get work. From the practice's perspective, good locum cover can be hard to find so a reliable, enthusiastic, hard working, newly qualified GP is just what they are looking for.

Should I sign a contract in advance?

This depends on who you are planning to work for. If you're planning to do sessions for a practice that you don't know and aren't sure what might be expected of you then it might be worth writing out a contract for you both to sign or at the very least giving them a copy of your own terms and conditions. To do this you will need to decide what you are willing to agree to in terms of rates, number of appointments, and visiting and detail it within a formal terms and conditions document, which you can then email to the practice along with your CV when you apply for the work. This way you avoid any awkward conversations about pay and what is expected of you and you can always renegotiate the contract after discussion with the practice manager if they require additional services. Once the surgery have agreed that they want you to do the work it is worth sending them a formal quote with a copy of your terms and conditions attached for both the practice manager and yourself to sign in advance of you doing the work. The other added benefit of this is that you can add in a section on cancellation fees and notice required so that practices can't book you and then bail out at the last minute. This may all sound a little ruthless but when you are self-employed and relying on specific work to pay the bills then every penny counts.

Some practices may issue you with their own terms and conditions contract in advance of any locum work. **Only** sign this if you are happy with what they are suggesting as there is no going back after the event! The likelihood is that they will put in clauses to prevent paying you for any extra work and the rates that they are suggesting may be considerably lower than what you would normally accept.

Tip: Before agreeing to do any locum work make sure that you know what is expected of you in terms of number of appointments, seeing extras, carrying out home visits, and checking admin such as post and blood results.

What if the locum practice has a different computer system from that which I am used to?

Unfortunately if this is the case then it is your responsibility to make sure that you learn how to use that system. Hopefully you will have taken my advice from Chapter 19 and made sure that you have had a go at using other computer packages in the run up to the end of your registrar year, but if you haven't, then there is normally someone at the practice that could give you a brief lesson if you try and organize something in advance. Turning up on the day that you start work having never used their computer system before will make you look daft and also make your surgery very difficult and slow!

There is some online guidance on how to use the different computer packages which you can find by doing a Google search but it's not the same as 'having a go' in real life and being shown how to do things quickly and efficiently.

How much locum work should I commit to, e.g. how many sessions per week?

Hmm… a difficult one. Essentially this is up to you but remember that as a GPST you are unlikely to have been doing more than eight sessions per week (by the time you have had your half-day release and private study half day—although this of course depends on which deanery you are training within). Take away also the time out that you have for tutorials and study leave and you'll find that in terms of session's worth of patient contact it's probably no more than seven. Working as a locum is tiring and it's very easy to bite off more sessions than you can chew if you're not careful. Once word gets around that you are reliable and available, in busy periods you may get three to four calls per day offering you work and you may find it difficult to say no. Remember that running yourself ragged is neither good for you nor the patients that you are seeing despite the attraction of a growing bank balance.

Tip: Running yourself ragged trying to do 10 sessions per week of locum work is neither good for you, nor the patients that you are seeing despite the attraction of a growing bank balance.

What documents will the employing practice need to see?

They are likely to request to see the following:

- GMC certificate
- MRCGP certificate
- Performers List confirmation (showing the correct role)
- Medical indemnity certificate (for the right type of work, e.g. GP locum and not salaried)
- Recent CRB certificate (this should have been done by the PCT when you applied to update your Performers List membership to fully qualified GP)
- Driving licence and car insurance certificate
- Proof of immunity to hepatitis B (if required to do exposure prone procedures you may also be asked to show evidence of your hepatitis C and HIV status)

It's worth scanning a copy of these documents onto a file within your computer, so that when requested you can either email or print and send them to the practice along with your terms and conditions plus your CV without too much bother. Some practice managers may request to see the original documents but most will be satisfied by a good quality copy.

Some locums will create themselves a website which details all of their CV information, patient and practice feedback, information regarding their rates and availability, plus a copy of their terms and conditions. Some websites also have the facility for practice managers to check up-to-date availability calendars and request bookings online but this is definitely not the norm so don't panic if the thought of having to do this makes you feel faint!

Will I need my own equipment?

You may be provided with most things that you need within the room that you are allocated to work in, although it is always better to have equipment which you know how to use and also know that it isn't broken! Make sure that before you start your session you have everything that you think you might need, as scrabbling around for a speculum in the middle of your surgery is likely to make you run behind and also looks unprofessional to the patients. You should not, however, rely on the practice providing you with the basic equipment as many practices will expect you to provide your own—it's therefore essential that you have your own well-stocked doctor's bag.

What will I need in my doctor's bag?

- Stethoscope
- Thermometer

- Peak flow meter
- Disposable gloves
- Urine and stool bottles
- Swabs
- FP10 (specific to the practice within which you are working)
- Fit notes
- Disposable peak flow mouth pieces
- Ophthalmoscope
- Otoscope
- Blood pressure cuff
- K-Y Jelly
- Tape measure

You may also want to carry some emergency drugs which you can either buy yourself from a chemist or obtain from the practice within which you are working.

The deal

How many appointments will I be given?

This depends on what you agree on in advance. Some surgeries will add on extras if required, so check their policy on this before you agree to anything. Normally you will be given two to three hours' worth of 10-minute appointments. Some practices may have breaks for coffee in the middle of their surgeries and others won't, so if you're someone that can't survive without a strong coffee and a HobNob mid morning then it's probably best to check this out before you agree to a three-hour surgery with no breaks!

Will I be expected to see extras?

You should only do this if agreed in advance. You don't want to be over the top and charge for one extra patient but if there are more than a few then you might want to charge extra for these. If, however, they've asked you to be duty doctor or on call for the day and are paying you as such, then seeing extras is likely to all be part of the deal.

Will I have to do home visits?

Yes but only if agreed in advance. If you are being paid to work a full day then most practices will include in this either one to two visits. If they need you to do more than this then you could consider charging extra (unless you are charging a large amount to start with!) but make it clear on the contract that you provide them with that this will be the case.

Will I have to look at and file laboratory results?

If you are doing odd days here and there then you probably won't be expected to do this. Not only is it difficult to comment on blood results requested by other clinicians but you also will not be around to follow-up abnormal results and make sure that they get actioned. Filing normal results may seem straightforward but unless the patient has a specific follow-up appointment with the clinician that requested the tests then things can often get missed. If you are working as a long-term locum within a practice then essentially you will be required to do all of the normal things that would be expected of you as a salaried doctor.

Will I have to deal with clinical letters and post?

In a similar way to blood results, if you are only doing one-off days here and there then it is unlikely that you will be expected to deal with any clinical post, However, if you are being paid to be the on-call or duty doctor for that day then you may be asked to deal with any urgent queries or clinical situations which arise from any post that day. If you are working as a more long-term locum for the practice then you are likely to be asked to deal with the post for whichever GP you are covering during your locum sessions. This often just involves making sure that there is nothing which urgently needs dealing with or coding and shouldn't be too much of a burden to you.

What about writing referrals?

This will depend on whether the practice has a secretary who types letters based on audio or Lexacom dictations. In some surgeries GPs create their own letters via the notes rather than creating dictations. Make sure if you are dictating letters that you have informed the relevant secretary and checked whether your dictation has properly recorded. The last thing you want is to be recalled back to a surgery to dictate a mislaid letter on a patient that you saw a few weeks ago.

> 'I always type my letters in the patient's notes and let the secretaries know where to find them. This prevents the problem of missing dictations or being accused of forgetting to do them at a later date!'
>
> *Locum GP*

What things shouldn't be expected of me as a locum?

This all depends on what you have agreed in advance. Many locums won't sign repeat prescriptions either because they don't want to take on the responsibility or maybe just because it is tedious and time consuming! You shouldn't really be expected to be responsible for a GP trainee (unless this has been agreed in advance) and you shouldn't have to clean the toilets. They can, however, ask you to do anything within reason, so if you have strong feelings about anything that you don't or won't do then you should probably discuss this in advance.

What is meant by locum etiquette?

You can look at this from two different angles, that of the locum and that of the employing practice. In brief, the practice will expect you to turn up on time, know how to work the computer, respect both the staff and the patients, leave the room that you use tidy, and not steal anything (or use their phone to ring your long lost aunty in Australia!). Remember that GP practices are run as a business and therefore anything that you use or consume will have been paid for by the partners. Saying that though, most of them are more than happy for you to help yourself to a cup of tea or coffee, but it probably isn't reasonable to eat your way through their entire biscuit stash.

From the view point of the locum, etiquette for the practice refers to them making you feel welcome, providing you with the required login details for the computer along with an adequate room in which to work, and adhering to the pre-agreed contract that you signed in terms of workload and what is expected of you. Very few surgeries will exploit locums and those that do will struggle to get good quality locum doctors in the future, as word spreads fast and no one likes being taken for a mug. However, the same is true of GPs and practice managers who via the 'jungle drums' will know who are the good and who are the bad locums to employ.

> 'We had a locum who had a high recall rate (i.e. patients bounced back to the partners), had odd prescribing habits, didn't use problem headings or our EMIS computer properly, was half an hour late on three occasions, and very disorganized. We gave her one of our rarely used feedback forms and told her that we did not wish her to do any other locums for us and have discussed her abilities with a few other local doctors. Her 'card' is marked locally. Remember, you all need to develop an excellent reputation. It leads to more work and partnerships and good salaried posts. Stay very professional.'
>
> *TPD and GP Partner*

Money stuff

How do I register as being self-employed? And when can I do it?

You need to register as 'self-employed' with the HMRC in order to continue to pay your national insurance contributions and complete your income tax self-assessment. Unfortunately you cannot do this in advance but it can be done by phone, post, or online, from the day that your self-employed work starts. Further information can be found on the HMRC website[1].

Tip: Make sure that you register as 'self-employed' within three months of starting as such, otherwise you could face a £100 fine from the HMRC.

How much should I charge as a locum?

Ah, the burning question on all locums' lips. This varies depending on how much you do, the way in which you split the charges for the day, what is required of you in terms of hours of patient contact and responsibility, and the area where you work. Asking those around you what they are charging often isn't helpful as for some reason people like to keep these sorts of things close to their heart for fear of competition.

There are many ways in which you can charge:

1 **Hourly rate** (per hour of patient appointments normally at 10-minute intervals)—the problem with this is that it doesn't take into account the admin work that is generated from seeing those patients and it makes it difficult if extras or phone calls get added to your list. You don't really want to cause trouble with your new locum practice by asking for extra money when they put a few simple phone calls at the end of your list, but at the same time it can get quite frustrating if they are consistently adding extra patients to your list without agreeing to pay you for them. At the end of the day, however, patient care comes first and I'm sure that you wouldn't refuse to see an extra sick patient just because the surgery were only paying you for the two hours' worth of patients that you had already seen. The most sensible thing to do is agree in advance what you will and won't charge for in these circumstances (Box 21.1); however, this can often be a hassle and fairly stressful. One way to get around it is to detail your position on this in your terms and conditions contract which you can send to the practice before agreeing to the locum work and get it signed. That way if there are any discrepancies

Box 21.1: Ideas of how to charge as a locum

Hourly rate (6 ×10-minute appointments per hour): £85

Visits: £25 each

Mileage: optional—most locums don't charge for this as you can claim the tax back on it, but if you are working for a rural practice it might deter them from sending you out to all of the nursing homes in the middle of nowhere!

you can always just refer back to this document. On this contract you should document how much you expect to get paid for work done over and above the hours of surgery time that you have done.

2 **Sessional rate**—a session is normally classed as four hours and ten minutes. This can include a combination of face-to-face appointments, telephone calls, post and bloods admin, signing prescriptions, and doing visits. Make sure that what they expect of you is not unreasonable to get done in the allocated session. This is probably an easier way of charging than via the hourly route although if you have only been asked to do a two-hour afternoon surgery, it will be far more cost-effective for the surgery to pay you by the hour.
3 **Daily rate**—this again depends on what is expected of you during that day and is something that you need to confirm with the employing practice in question. If you normally charge £425 as a daily rate for two two-hour surgeries plus two visits and admin, you don't want to charge the same rate for a practice that expects you to do two three-hour surgeries, five visits, be responsible for a registrar, and on call covering two surgeries! Check this out in advance so that you don't get stung!

Essentially what you need to remember is that by working as a locum you are getting yourself known in the local area for the first time as a proper GP and the last thing you want to do is get yourself a bad reputation from the start. If you charge reasonable rates and work hard then locum work will often lead to salaried post opportunities or even partnerships. If you are lazy, obstructive, and charge the earth then the chances are that not only will you get very little locum work but it might also jeopardize your chances when applying for more permanent positions. Remember—practice managers within each area meet on a regular basis and they talk about these sorts of things.

> 'Pay is important but negotiate up front, try to be fair, and the practice should treat you fairly.'
>
> *TPD and GP Partner*

Do I need an accountant?

This all depends on how organized you are and whether or not you understand accounting and tax returns. Many newly qualified locums will perhaps get an accountant to help them complete their first year's tax return and from that point onwards can manage without. If you are one of those people that want to make sure that you have claimed back every possible penny in tax relief or would like to invest your money in weird and wonderful things, then an accountant is definitely for you. Depending on how much you want them to do their fees will vary, so don't forget to calculate this into the equation when working out what's most financially viable. If you are good with spreadsheets (and have a love for Excel) and are extremely organized then you can probably do most of it yourself very simply without having to pay anyone for the pleasure. Most people, however, end up doing something in between and pay just a small amount such as £200 for an accountant to check over their tax return once a year and make sure that they're not paying out more than they should.

How do I continue to pay my National Insurance?

There are many different types of National Insurance and it can get quite confusing if you've never really had to think about it before. As a PAYE employee (which you will have been during your training and hospital years) your National Insurance contributions (NICs) will have been automatically taken from your salary, the amount of which will be detailed on your payslip. Once you become self-employed and have registered as such with the Inland Revenue you will be required to pay two different types of National Insurance:

- Class 2
- Class 4

Class 2 National Insurance is at present £2.40 per week and should be paid by direct debit on a monthly basis as soon as you become self-employed. This amount is fixed regardless of how much or how little income you earn. Like tax however, there is a certain threshold below which you will not have to pay any Class 2 National Insurance. In 2009/10 this was £5,075 per year. This is particularly relevant for those who are doing a small amount of extra locum work as a self-employed doctor. If, however, you are earning more than this per year then you will need to pay both Class 2 and Class 4 NICs (the threshold for Class 4 contributions in 2009/10 was £5,715 per year).

Class 4 National Insurance, like your tax, is dependent on the amount of **profit** that you make from your self-employed work. This will be calculated when you complete your tax return but at present is set at 8% of any profits up to a certain value (£43,875 in 2009/10) and 1% for profits above that (Box 21.2).

Box 21.2: Worked example for Class 4 NICs (based on 2009/10 figures)

Doctor Foster earned £55,000 gross income during 2009/10

His expenses totalled £5,410

His **profit** therefore that year was £49,590

Class 4 contribution calculation (2009/10 = 8% on profits between £5,715–£43,875 and 1% on profits above £43,875):

£49,590 – £5,715 = £43,875

£43,875 × 8% = **£3510**

If after deducting expenses Dr Foster had earned **more** than the upper Class 4 limit then the extra profit would be charged at 1%. If maths never was your strong point then you might be one of those that would benefit from paying an accountant;

however, your self-assessment tax return will calculate it for you at the end of the year so don't get too stressed by it all. Essentially, if you budget that about 8% of your earnings will need to be paid in National Insurance at the end of the year then you can't go too far wrong.

Am I still eligible for an NHS pension?

To continue to contribute towards your NHS pension as a locum, you will need to submit a GP Locum Form A and GP Locum Form B for each period of work that you do. These forms are available online via the NHS pension's website[2]. Remember, however, that not all locum work is NHS pensionable; for example, some OOH providers and all locum agency work is not. This may affect whether you decide to work for them.

Locum Form A is a certificate which confirms that you have completed a period of GMS/PMS or APMS GP locum NHS work (and therefore cannot be used to record any work done 'out of hours' or for a non-GMS/PMS/APMS practice). Details of how to complete it can be found on the second page of the form but it is essentially very straightforward. You will need to submit a Locum A Form to the PCT for each **month** and for each **practice** within which you work. Each form needs to be signed by the employing practice to confirm how much money you have been paid for the work. This is normally done by the practice manager but it is your responsibility to make sure that it is completed and submitted to the PCT within the required time period. The easiest way to get this done is to complete it and attach it to the invoice that you send to the practice at the end of each period of work.

Locum Form B is a record of all of the locum work you have done within each month and is used to calculate the amount of pension contributions that you are required to make to the PCT. The tiered contribution rate that you are required to pay depends on the gross amount of pay that you earned in the previous pension year (that is April to March rather than per calendar year). There are four tiers of contribution rate which in 2009/10 were based on the following gross levels of pay:

5%: up to £20,224
6.5%: £20,225–£66,789 (this is the rate which most of you will be paying)
7.5%: £66,790–£105,318
8.5%: £105,319 plus

The salary ranges for each tier of contributions is based on full-time equivalents, so if you are working part time then you will need to factor this in to your calculation. Once you have decided which contribution rate you are required to pay, you will then be able to calculate the total amount of NHS Pension Scheme employee contributions that you are required to make for that month's locum work. There is, however, a 10% allowance for your professional expenses so you will only end up paying a percentage of 90% of your total earnings for the month.

Whilst this may initially seem complicated, once you get used to doing it, it really is quite straightforward (Box 21.3).

Box 21.3: How to calculate your NHS pension contributions

Dr Foster earns £4,000 for locum work done in April
10% of this is deducted for professional expenses leaving £3,600
NHS pension contributions at 8% of this value = £3,600 × 0.08 = £288

There may be adjustments that you need to make to this if you have paid any extra pension contributions over the years but this is the basic calculation. Once calculated you will need to write a cheque for the full amount made payable to your local PCT and post it along with the Locum B form and all Locum A forms for that month. Make sure that you make copies of all Locum A and B forms that you complete before sending them and also keep a note of the amount of pension contributions that you make each month. It is also wise to request that the PCT send you a receipt for the amount of pension that you have paid so that you can use this when working out your tax return.

Tip: Don't forget to ask the PCT to send you a receipt for the pension payments that you make as these will be required when you submit your tax return.

If all of the above sounds incredibly daunting and you'd rather have the forms all sorted for you then there is a website called http://www.pennyperfect.com which is designed to help GP locums organize their finances and complete the required forms. There is a charge to register with the site but it is minimal and is felt to be well worth it by those locums who choose to use it.

It is also worth mentioning that if you are a session-based GP engaged on a long-term basis then you are no longer classified as a GP locum and instead are considered to be a type 2 medical practitioner (GP performer). In this case it is the practice that should be responsible for paying the employer contributions and not the PCT, which may affect how you organize your own payments. Further information on this can be found on the NHS pension's website[2].

Will I still be entitled to an NHS pension?

Yes, as long as you continue to make earnings related contributions via the PCT as detailed in the previous question.

How much tax will I have to pay and when?

Once you have registered with HMRC as being self-employed, you will then be expected to complete a self-assessment tax return for each tax year. You should receive a letter at the end of each tax year asking you to complete a tax return either by post or online.

If you want to complete your tax return online then you will need to register via the HMRC website[3].

Once you have registered with the HMRC as being 'self-employed' (which you will need to do if you are planning to work as a locum) you will be sent a letter confirming that you will be required to submit a tax return each year. This self assessment is used to calculate how much income tax, Class 4 NICs, and student loan contributions (if still outstanding) you will be required to pay for that tax year. The tax return is submitted retrospectively so you will need to make sure that you have set aside a pot of money to make this payment at the end of the tax year.

How do I submit a tax return?

Tax returns can be submitted either by post or online. If you wish to complete it online then you will need to enrol and register for this via the HMRC website[4]. To do this you will need your Unique Taxpayer Reference (UTR) which you should be able to find on any paperwork from the HMRC. Once you have filled in your personal details via the website, you will be allocated a user ID which you will need to access the online HMRC facilities. The user ID and required activation code should then be sent to you by post within the next seven days. Using these codes you will then be able to register for the 'Government Gateway' which enables you to register for the multiple online e-Government services, of which online tax return submission is one. You normally have until the end of January the following year to submit your tax return but if you decide to do it by post then the deadline is much earlier, being the end of October that year.

How much tax will I pay?

This depends on how much you have earned and is essentially the same as if you were employed and getting paid via PAYE. Based on 2009/10 figures you will pay 20% on the first £43,875 earned and then 40% on any earnings above that. Don't forget, however, that you get a tax free allowance of £6,475 so remember to take this into consideration when you're working out your sums. Also bear in mind that the tax that you pay is based on the profit that you have made and therefore your professional expenses will be taken into account when calculating the amount of tax owed.

If you decide to submit your tax return online then you will get an instant figure for the amount of tax, NICs, and student loan contributions that you owe. Although you normally have until the end of January the following year to submit the return, doing it early means that you have more of an idea where you stand financially.

What type of expenses can I claim back for tax purposes?

This is really a question for an accountant but essentially once self-employed you can claim for anything that is necessary in order for you to carry out your work. This includes travel expenses, professional subscriptions or fees, mobile phone or Internet costs, postage and printing costs, medical indemnity insurance, etc. You need to be honest about it though and keep meticulous records; otherwise you could find yourself in deep water with the tax man. Remember also that if you are both employed **and** self employed, then you can only claim back those extra expenses for your self-employed work and not for your employed work. In the eyes of the tax man, your employer should be reimbursing you for any essential work-related expenses and so you need to be careful on this one.

Should I get any income insurance?

This depends on how much you like taking a risk. Essentially it comes down to the simple question 'how will you survive financially if you can't work for 'x' amount of weeks?'. Most people could survive in the short term (e.g. a few weeks) but if you break a leg or develop a moderately serious illness which prevents you from working for a few months, then you might struggle. Many companies provide several types of sickness insurance for those who are self-employed so that in the instance of short- to long-term sickness, you will not be left without any income. Other policies such as income protection and critical illness may also be an advantage although this is also relevant to those in employed posts. It is possible to get 'unemployment' insurance too, which kicks in if for some reason it is impossible for you to get any work. This is unlikely to be a fruitful option for you, as you will need to provide good evidence that there is definitely no work available, which for a qualified doctor is quite unlikely (and the insurance companies know this).

How do I invoice the practice that I am working for?

Create your own headed standard invoice template using Microsoft Word so that you can simply adapt it to become relevant to the work that you are invoicing for. Think about including the following:

- Make it headed, e.g. your name and titles across the top
- Date of invoice
- Date of work done
- Type of work done, e.g. one morning session or full on-call day
- Amount you are charging for that work
- Total amount payable by the practice
- Statement about when the payment is due

If you have registered with http://www.pennyperfect.co.uk then this website can also be used to create your invoices.

> Tip: Make sure that you keep a hardcopy of each invoice sent so that should your computer decide to crash one day, you can still submit your tax return without having too many palpitations. You may also want to encrypt your hard drive to protect your financial information.

When will they pay me?

This depends on what you put in your invoice and how efficient the practice manager is. Be prepared to do some chasing as the chances are that the person who needs to sign the cheque/authorize the payment is not around (hence why you're doing the locum!). Most practice managers are pretty efficient when it comes to making locum payments but there are others who may bury your invoice under a pile of QOF data and only realize that you haven't been paid when you send them a begging email. It's not unreasonable to expect to be paid within 14 days of doing the work and in fact some locums will demand that they are paid within seven days. If you provide them with your internet banking details then the practice are able to pay you by BACS transfer rather than writing you a cheque (which you would then have to pay in at the bank and may take a few days to clear). As it gets nearer to the end of the month it becomes more important that you get paid on time so that you can complete the other necessary forms and paperwork. Compared to being paid via PAYE, it can be relatively stressful chasing practice managers for money, but if you are organized and pleasant about it then it shouldn't be too much of a problem. Remember though that if you are invoicing a practice at the end of each month then it's likely to be at least a week before that money is available within your account, which may have ramifications if your mortgage and other bills come out of your bank within the first few days of the month. Make sure that you have a buffer of money within your account to make sure that you don't end up with lots of nasty overdraft charges.

> Tip: Keep a record of the date that your invoices are sent and the date that payment is received. Giving each of your invoices a unique ID number may help when it comes to submitting your tax return.

Special circumstances

Can I get maternity pay as a locum?

Yes, but it is likely to be paid in the form of maternity allowance (MA) rather than statutory maternity pay (SMP). This can get very complicated depending on your situation but if you have been paying your NICs like you should then you should be entitled to MA. This is calculated as a percentage of your average salary over the preceding three months up to a maximum amount of around £500 per month. As a locum you are unlikely to get any occupational maternity pay (e.g. paid to you by your employer) and instead the MA that you receive is paid to you by the government and is therefore at no cost to the practice in which you work. More information on maternity pay can be found in **CHAPTER 15, 'LIFE EVENTS AND CHANGES THAT MAY AFFECT YOUR TRAINING'**. You will not get any other paid leave as a locum unless agreed in advance or covered by a specific insurance policy.

What happens if they cancel my work?

This depends on whether or not you have signed an agreement in advance. Some locums will put a clause in their contract which requires the practice to pay a specific charge if the work is cancelled within a certain time period, but essentially it depends on how you want to run your 'business' as it were. If you are doing the odd locum here and there on top of a regular job then you may not be as worried about cancellation of work. If, however, you are relying on the work to pay the mortgage (and have turned down alternative work because you were already booked) then you are likely to be less impressed with the work being cancelled.

What happens if I am sick and can't go to work?

Unfortunately there isn't much that you can do about this one. If you're genuinely ill then you shouldn't work. Whilst the prospect of missing out on that day's or week's pay might be worrying, you need to make sure that you are looking after yourself. Remember the old saying 'you need to look after yourself before you can look after others'.

There are some insurance policies that you can take out to cover short-term sickness as a self-employed individual, but they don't tend to pay out until after one month of illness. Longer-term policies to cover illness such as critical illness protection are also available but tend to be very specific on what illnesses they cover.

Can I still do locum work if I have a salaried job?

Yes you can but if you earn any money in addition to your regular salary then you will need to declare this by completing a self-assessment tax return at the end of each tax year. If you are planning to earn more than around £5,000 you will also need to register as being 'self-employed' in addition to being 'employed', as you may be required to pay additional national insurance contributions.

Where can I go for further information?

Should you require any further information or feel that you would benefit from the support of a big organization then think about joining the National Association of Sessional GPs at http://www.nasgp.org.uk. It'll cost you around £8 per month to join and you can register either online or by downloading the registration form and sending it by post.

REFERENCES

1 HM Revenue and Customs. *Register as self-employed*. Available at: http://www.hmrc.gov.uk/selfemployed/iwtregister-as-self-employed.htm (accessed 14 August 2010).

2 NHS Pensions. Member Forms. Available at: http://www.nhsbsa.nhs.uk/MemberForms.aspx (accessed 14 August 2010).

3 HM Revenue and Customs. *Understanding and using self-assessment online*. Available at: http://www.hmrc.gov.uk/sa/understand-online.htm (accessed 14 August 2010).

4 HM Revenue and Customs. *Welcome to HMCR online services*. Available at: https://online.hmrc.gov.uk/login (accessed 14 August 2010).

PART 5

Post VTS

CHAPTER 22

Life after the VTS (and where to go for support after leaving your trainer!)

The first few months after completing the VTS can be a scary time for newly qualified GPs, regardless of whether you're working as a salaried GP, a locum, or even your trial period as a partner. You are all of a sudden expected to practise independently and autonomously without the supportive crutch of your trainer in the room next door. You also no longer have the weekly VTS teaching to look forward to, which whilst you may not have appreciated it at the time, was an essential venting opportunity, not to mention a social and educational one whereby you got to spend half a day with your colleagues in a fairly relaxed, informal setting.

Starting out on your own can make you feel vulnerable and isolated, particularly if you are still working as a locum or have joined a practice or area which was not familiar to you. Make sure that you stay in touch with your colleagues and if possible try to meet with them on a regular basis.

Who can I turn to for support once I've completed my training?

There are many different support networks to which you can turn, the most important one being your peers and colleagues that you have spent the last three years training alongside. As mentioned previously there will also be your trainer, educational supervisor, and course organizers who will be more than willing to help you along your way in those first few months.

The RCGP have also recently developed an initiative entitled 'First 5' which was created by the RCGP AiTs in order to provide support to newly qualified GPs from completion of training until the point of revalidation five years later. It was developed after the recognition that whilst at the end of ST3 AiTs may be competent in the required elements of the RCGP curriculum, they may, however, still lack confidence in certain areas and may feel either vulnerable or isolated as a new independent practitioner. It plans to organize courses on topics such as practice management, leadership, and advanced consulting which were felt by newly qualified GPs to be the subject areas where knowledge was perhaps lacking.

It was also hoped that First 5 would encourage members to maintain their membership with the RCGP as the benefits of doing so are not always obvious to the newly qualified GP. As a fairly new addition to the college there are bound to be many more opportunities on offer in the future via this initiative.

Can I continue to use my ePortfolio?

Yes—although with the introduction of revalidation, as of autumn 2010 you will be required to use an alternative ePortfolio to log all of your CPD activity. Watch this space on that one!

Are there any groups that I can join?

There are plenty of groups both locally and nationally where you can get further support and advice on work, contracts, and CPD activities. You will probably find that there will be a local salaried GPs group which will provide not only regular educational meetings but if you join their mailing list will also email you with local salaried and locum opportunities, before they are published in the national press. You may also want to consider forming your own small group with your peers from the VTS even if it's just to organize a monthly dinner out rather than an educational meeting.

If you wish to join a more comprehensive group then consider paying to be part of the National Association of Sessional GPs (NASGP)[1]. Whilst this will cost you a small amount to join, it promotes itself as being the 'only independent lobbying and information service for Sessional GPs'.

What's the best way of finding out about forthcoming jobs?

Network, network, and network. Many job opportunities whilst formally advertised may be already decided upon based on local hearsay and word of mouth. Make sure that you are in the right place at the right time by frequenting local sessional GP meetings, attending educational events, and participating in local training days. Stay in touch with other recently qualified VTS trainees who may well have just joined a practice who are looking for an additional GP. And if all of that fails, then keep your eye on *BMJ* careers and keep searching!

REFERENCE

1 National Association of Sessional GPs (NASGP). Available at: http://www.nasgp.org.uk (accessed 14 August 2010).

CHAPTER 23

Revalidation and appraisal

Whilst it seems a long way off when you're only at the start of your GP training, whether you like it or not, revalidation for all qualified GPs is planned, with the prediction that by 2016, all GPs within the UK will have been assessed. Appraisal, however, has been around for a long time and will continue to be integral to the new revalidation process rather than be replaced by it.

Appraisal

What is appraisal?

The appraisal interview is very similar to the educational supervisor reviews that you will have completed during your ST years. It occurs annually and is a formal process by which your professional commitments, achievements, and clinical practice are evaluated by a designated local appraiser. As part of the appraisal process you are required to gather a fairly large amount of information based on the work that you have done during that appraisal year. This was previously all collated via the NHS appraisal toolkit; however, as of autumn 2010 it is due to be replaced by the new revalidation ePortfolio. This, however, is still in its infancy and therefore there are likely to be many changes to how this is organized. Whichever method is used, there are several different sections which you will need to complete in preparation for your appraisal and it can take a considerable amount of time depending on how much you have achieved in the preceding appraisal year. You should allow sufficient time in order to complete it accurately. The appraisal year runs from 1 April to 31 March regardless of when you qualify, so if you don't get one done by your trainer before you finish then you will need to get one completed before the end of the appraisal year in your new post. This may be a little frantic so if you are able to get your trainer to carry out an appraisal and complete all of the necessary forms before you finish your training then it gives you a little grace. Participating in the annual NHS GP appraisal forms part of the compulsory Performers List regulations and therefore if you have not completed this within the required time, you may be at risk of losing your place on the Performers List.

Who will be my appraiser?

This is normally organized by your local PCT who will usually give you a choice of a few registered appraisers within your area. Find out early on who you need to contact at the PCT and when with regard to organizing this so that you're not racing around trying to arrange it at the last minute. It is your responsibility to contact your appraiser and organize the meeting in a similar way to your educational supervisors' reviews. Once you have chosen your appraiser they are then able to access your documents via the toolkit so that they can prepare in advance for the appraisal (or via your ePortfolio once these become active).

Revalidation

What's the difference between relicensing and recertification?

At present, all fully qualified GPs are required to be certificated on the GMC GP register and hold a 'licence' to practice.

LICENSING

This came into force in 2009 and is now a requirement by law for all practising doctors regardless of their grade and the type of work that they do. Only doctors with a licence to practice are able work as a doctor in the NHS, write prescriptions, and sign death or cremation forms. Licensing is just the first step in the process of revalidation and will depend on periodic renewal.

CERTIFICATION

Doctors on the specialist or GP register are also required to recertify against the standards which have been set by the relevant college or faculty and approved by the GMC.

REVALIDATION

Revalidation, however, is an amalgamation of the two processes of relicensing and recertification and is currently still work in progress. It will nevertheless offer us as GPs the opportunity to confirm to the general public that we are up to date with what we should know and are fit to practise. But it isn't just about restoring the faith in our profession (although that would be nice!), it's a process that will allow us to reflect and improve on the care that we deliver, in a similar way to the GPST WPBA.

What information will I need to collate for revalidation?

The information that you collate for your annual appraisal will form the basis of your revalidation material, which you will then be required to submit online via an ePortfolio which is very similar to that which you used as an AiT. Information which you will be required to submit will include the following:

- A description of your current work
- Any special circumstances which may affect (or have affected) your work, e.g. sickness or maternity leave
- Evidence of annual appraisal 'sign off'
- PDPs and the subsequent reviews of these completed at your appraisal
- Evidence of your CPD achievements—it has been suggested that you should be accumulating at least 50 credits per year
- Multi-source feedback reports
- Patient satisfaction questionnaires
- Reviews of any formal complaints made against you
- Significant event audits
- Audits looking at your own practice and care of patients
- Statement of health and probity

Look familiar?

Essentially the appraisal and revalidation requirements are very similar to those which you have been used to completing during your training years, the only difference being that you will now be responsible for making sure that you are keeping your records up to date and will not be prompted by a forthcoming panel review to make sure you have submitted the right amount of DOPS etc. The RCGP are also in the process of developing a learning credit system whereby you accumulate credits for various different CPD activities that you manage to achieve. The plan is to steer away from the 'one credit per hour' idea and instead base the number of credits achieved not only on the time taken for that activity but also on the degree of improvement in care or practice that may result from it.

What is the new revalidation ePortfolio?

All RCGP members will be given a revalidation ePortfolio as part of their subscription and it is hoped that it will contain a learning log similar to that in the AiT ePortfolio as well as a traffic light system to indicate your progression towards the revalidation process. There will also be direct links to RCGP e-learning modules which can then be recorded within it. This should all come as no great stress to newly qualified GPSTs who are all too familiar with completing online assessments and logging educational activities and achievements. It is hoped that there will be a smooth transition for trainees moving from their AiT ePortfolio to the revalidation ePortfolio but as with all new initiatives there are likely to be plenty of teething problems.

How often will revalidation take place?

Revalidation will be repeated on a five-yearly basis and will not replace the annual appraisal that you will still be required to complete at a local level.

How much does it cost to get a licence?

This is included in your annual GMC registration fee if you are paying the full amount for registration with a licence to practice. In April 2010 this was £420 and is actually not any different to the fee that was payable prior to relicensing coming into force in 2009 (the only slight rise compared with 2009 was due to inflation). If you wish to register or retain your registration without a licence to practice then you will pay the reduced rate of £145, although without a licence you will be unable to do any clinical work. The GMC website gives very clear advice on what you can and can't do without a licence.

What if I am a locum?

All practising GPs will be required to revalidate regardless of the type of work that they do. As a locum it may be slightly more difficult to collate all the information that you need, particularly if you are working in lots of different practices. Make sure that you allow enough time within the working week to think about what you need to be accumulating for your appraisal and revalidation, as unlike with a salaried post or a partnership position there will not be any designated time allocated for this and instead it will depend on your own motivation and inclination. Working as a locum for 10 sessions a week is fine if you can stomach it, but don't forget that you will need to be spending some time on your continued professional development, for example, by attending courses, carrying out audits of your practice, and completing significant event audits.

What is self-directed learning (SDL)?

This refers to anything which you have chosen to partake in yourself and forms an important part of appraisal and revalidation. Some examples of SDL are:

- Creating a reflective diary of consultations
- Log of referrals and investigations
- Reviewing the outcomes of referrals

- Reading journals/textbooks
- Completing online e-modules

Where can I go for more information on revalidation?

All of the essential information that you need on revalidation can be found on the GMC website which contains the answers to a significant number of frequently asked questions. The RCGP website and newspaper also contains regular updates on the plans for revalidation so it's worth making sure you keep up to speed with the changes taking place.

CHAPTER 24

If your question isn't answered in any of the above...

Whilst this book aims to have considered most of the questions that you may have had either prior to commencing or during your GP training years, it would be impossible to cover everything and therefore there is likely to still be information that you need.

If so, here are some of the contact details of organizations that may be able to help in providing you with any extra information that you require. And if you still can't find what you're looking for... ask your trainer!

Professional contacts

BMA:	http://www.bma.org.uk
GMC:	http://www.gmc-uk.org
GMC certification registration:	http://www.gmc-uk.org/certificationonline
RCGP:	http://www.rcgp.org.uk

Medical Defence Organizations

MDDUS:	http://www.mddus.com
MDU:	http://www.the-mdu.com
MPS:	http://www.medicalprotection.org.uk

Educational stuff

GPNotebook:	http://www.gpnotebook.co.uk
Primary Care Forms:	http://www.primarycareforms.com

Jobs stuff

BMJ Careers:	http://www.bmjcareers.co.uk
Medicins Sans Frontiers:	http://www.msf.org.uk
NHS Jobs:	http://www.jobs.nhs.uk

Money stuff

Inland Revenue:	http://www.hmrc.gov.uk
NHS employers:	http://www.nhsemployers.org

Locum stuff

Locum organizer:	http://www.locumorganiser.com
NASGP:	http://www.nasgp.org.uk
Penny perfect:	http://www.pennyperfect.com

Useful email addresses

AiT helpline:	ait@rcgp.org.uk
BMA:	support@bma.org.uk
RCGP exams:	exams@rcgp.org.uk

PART 6

Additional Content

CHAPTER 25

The coffee break quiz—time for a few practice AKT questions!

If you're intrigued about what types of questions you may be faced with in AKT then read on below to find a selection of examples for you to try your hand at. Remember that the questions are produced in a variety of different formats and it is therefore important that you familiarise yourself with plenty of questions in preparation for the exam (and by plenty I mean hundreds!). Whilst these examples are probably not of the exact complexity level that you would expect in the exam, they will give you a taste of what is to come and the types of phrasing used within the questions. All answers can be found at the end of the chapter.

Curriculum statement 3.1—clinical governance

1 The QIPP programme is a new government initiative, which ensures that each pound of NHS money spent is used to bring maximum quality of care and benefit to the patients. Which one of the following correctly translates the acronym QIPP?

A Quality, Innovation, Performance, and Presentation
B Quality, Innovation, Productivity, and Prevention
C Quality, Initiation, Productivity, and Prevention
D Quality, Innovation, Performance, and Prevention
E Questioning Innovation, Productivity, and Prevention

Curriculum statement 3.5—evidence-based practice

2 A 25-year-old girl comes to discuss her smear results with you which have shown the presence of cervical intra-epithelial neoplasia (CIN3). She has read on the Internet that primary cervical screening carries a 50–70% sensitivity to detect CIN3. How do you explain this to her in lay terms? Please choose the single most appropriate answer from the list below.

A 50–70% of women with CIN3 will have an abnormal smear result

B 30–50% of women with CIN3 will have an abnormal smear result

C 50–70% of women with an abnormal smear result will have CIN3

D 50–70% of women with a normal smear result will have CIN3

E 30–50% of women with CIN3 will have a normal smear result

Curriculum statement 7.0—care of acutely ill people

3 Jasmine, aged 21, presents as an urgent extra to your already busy morning surgery. She describes a 24-hour history of progressive right-sided abdominal pain associated with nausea, vomiting, and increased vaginal discharge. Which of the following tests is most likely to provide a definitive diagnosis?

A Temperature

B Urine pregnancy test

C Urine dipstick

D High vaginal and *Chlamydia* swab

E Abdominal and pelvic examination

Curriculum statement 10.1—women's health

4 Ms Padwick, a 53-year-old lawyer, comes to see you concerned about cervical cancer. She is asymptomatic and last had a smear when she was 51. She has never had an abnormal smear. When is her next smear due as part of the national screening programme? Please choose the single most appropriate answer from the list below:

A Age 58

B Age 57

C Age 54

D Age 56

E She no longer requires smear tests.

Curriculum statement 10.1—women's heath

5 53-year-old Mrs Haynes presents to you at the end of your surgery one night in despair about her hot flushes, which she has been struggling with for the last 2 years. She has been amenorrhoeic for nearly 18 months and is desperate to try HRT after having had no benefit from any of the 'over the counter' preparations for menopausal symptoms. She is otherwise fit and well with no significant medical history. After careful counselling and examination, which one of the following medications would you be most likely to prescribe?

A Elleste-Solo 1mg OD

B Tibolone

C Elleste-Duet 1mg OD

D Elleste – Duet Conti

E Evorel Patch '25'

Curriculum statement 10.2—men's health

6 29-year-old Mr Hall presents to you with a copy of a semen analysis that he recently had performed privately. It shows severe teratoazoospermia and he would like to be referred to the local NHS fertility clinic, as he and his wife have been trying for a baby for the last 6 months with no success. Which one of the following options would be the least appropriate course of action?

A Repeat the sperm analysis via the NHS in 1 month

B Advise Mr Hall to reduce his caffeine intake and stop smoking

C Advise Mr Hall that she should maintain a healthy diet and engage in some regular exercise

D Recommend that he and his wife try to have regular intercourse every 2–3 days

E Refer to the local NHS Centre for Reproductive Medicine (CRM) clinic

Curriculum statement 11.0—sexual health

7 26-year-old Natalie presents complaining that her boyfriend has noticed small lumps around her vagina. She denies any associated discomfort and has not noticed them before. Her last smear test was normal 6 months ago. On examination, it becomes apparent that she has genital warts. Which one of the following would not be an appropriate part of her management plan?

- **A** Advise her to use condoms during intercourse in order to prevent transmission and re-exposure to the virus
- **B** Inform her that the majority of patients with human papillomavirus (HPV) clear the virus within 12 months
- **C** Given that HPV infection increases your risk of cervical cancer she must have an additional cervical smear performed
- **D** Screening for other sexually-transmitted infections (STIs)
- **E** Treatment options vary depending on the type, number and distribution of the warts

Curriculum statement 13.0—care of people with mental health problems

In each elderly patient below match the most likely diagnosis to the presentation. Each option may be used once, more than once, or not at all.

- **A** Appropriate response for circumstances
- **B** Delirium
- **C** Dementia
- **D** Delirium and dementia
- **E** Depression

Patients

8 A 74-year-old lady who presents with her daughter who is concerned that her mother hasn't eaten for over a week since the sudden death of her husband and was yesterday found wandering down the road in her pyjamas.

9 An 85-year-old lady from the local nursing home who was found pulling up daffodils in the garden looking for her slippers with a background of a 3-day history of urinary incontinence.

10 A 90-year-old lady, who is the main carer for her disabled husband, presents with a 6-week history of poor memory, anorexia, and feeling tired all the time.

Curriculum statement 10.2—men's health

11 52-year-old builder, John, presents to you concerned about prostate cancer after reading an article in the Daily Mail. He is concerned that his uncle died from prostate cancer at the age of 83 and is asking for a prostate-specific antigen (PSA) test. He isn't keen to be examined today as he has come straight from work. Which one of the following management options would be most appropriate?

A Reassure him that prostate cancer is very common in elderly men and that no further tests are required

B After obtaining his consent, examine his prostate and request a PSA blood test the next day

C Offer to carry out a PSA blood test today, but explain that he will need to return for a rectal examination to assess the size and consistency of his prostate

D Request an USS of his prostate to assess size and consistency.

E Offer to carry out a PSA blood test and ask him to call back for the results in 1 week. If normal then no further action is required.

Curriculum statement 15.1—cardiovascular problems; 15.8—respiratory problems

The following three patients all present during your morning surgery complaining of chest pain. In each case, match the most likely diagnosis to the presentation. Each option may be used once, more than once, or not at all.

Options:

A Hyperventilation and anxiety

B Myocardial infarction

C Oesophageal spasm

D Chest infection

E Pneumothorax

12 A 16-year-old girl on the COC (who is in the middle of sitting her GCSE exams) presents in tears with a 2-hour history of bilateral chest tightness associated with rapid breathing. Clinical examination is unremarkable and her saturations are 100% on air. She has no other risk factors for VTE (venous thrombo-embolism) and says that she had a similar episode last week, which settled after deep breathing.

13 A 56-year-old smoker presents acutely with a 30-minute history of crushing central chest pain, which started after eating steak and chips in the pub. He says that he has been getting episodes of 'indigestion' on and off for the last few weeks, but has been ignoring them. He has tried Gaviscon with no relief and he is becoming a little breathless.

14 Jamie, a 19-year-old semi-professional footballer presents one evening saying that he became acutely breathless during training tonight and has noticed a sharp pain in the right-hand side of his chest when he breathes in. He has had a cough for the last few days, but denies any fever or sputum production. Examination reveals decreased vocal fremitus on the right hand side.

Curriculum statement 15.2—digestive problems

15 45-year-old Mandy presents with a 6-month history of abdominal cramps, bloating, alternating bowel habit, and some loss of appetite. Her mother died of bowel cancer 7 months ago and she has been struggling to come to terms with her death. Examination is entirely normal. Which one of the following would be the most appropriate management plan?

A Reassure her that she doesn't have bowel cancer

B Request full blood count (FBC), thyroid-stimulating hormone (TSH), and C-reactive protein (CRP), and organize an urgent colonoscopy.

C Discuss the possibility of irritable bowel syndrome (IBS) and prescribe a course of Colofac and Fybogel.

D Request an urgent abdominal X-ray

E Refer her to the colorectal clinic for genetic screening

Curriculum statement 15.3—drug and alcohol problems

16 You have been asked by reception to see a temporary resident who has presented to the front desk requesting an urgent appointment to see a doctor. He is staying in the area for the next 3 days and has left his medication back at home 300 miles away. He is not known to the practice, but tells the receptionist that he has poorly-controlled epilepsy and may suffer a seizure if he doesn't get hold of his medication today. You agree to see him at the end of surgery when he requests tramadol 50mg 100 tablets, diazepam 2mg 28 tablets, epilim 500mg 56 tablets and paracetamol 500mg 100 tablets. What would be the next most appropriate course of action?

A Refuse to prescribe any medication

B Ask the patient to return the next day once you have contacted his regular surgery to confirm his regular medication

C Prescribe the requested medication.

D Prescribe a 7-day supply of the requested medications

E Prescribe a 1-day supply of medication and ask him to return the next day once you have been able to contact his usual GP

Curriculum statement 15.4—ENT and facial problems

17 3-year-old Joshua attends your morning surgery complaining of right-sided earache for the last 24 hours. He is systemically well. Examination reveals a temperature of 37.5°C, a clear external auditory canal, but the presence of a bulging right tympanic membrane. Which of the following would not be an appropriate management option?

- **A** Advise the parents regarding regular paracetamol and 'wait and see'
- **B** Prescribe a standby script for 5 days of amoxicillin to be used if things worsen over the next 72 hours
- **C** Prescribe topical antibiotics drops and a standby script for amoxicillin if things don't improve
- **D** Suggest regular ibuprofen and review the next day if not improving
- **E** Advise the parents that his eardrum may burst over the next few days

Curriculum statement 15.6—metabolic problems

18 Your next patient, 27-year-old Neil Parker, comes to see you regarding his recent blood results, which have shown a TSH of <0.01 and a free T4 of 69.4. You make a diagnosis of thyrotoxicosis and decide to commence him on carbimazole, taking care to counsel him about the possibility of agranulocytosis. Which one of the following would you advise him to report immediately once he has started the treatment?

- **A** Mouth ulcers
- **B** Sore throat
- **C** Bruising
- **D** Fever
- **E** Any of the above

Curriculum statement 15.7—neurological problems

19 32-year-old Jayne Hawkins comes to see you regarding the increased frequency of her migraines. She admits that she has been very stressed and anxious recently and, as a result, is often having at least two migraine attacks per month, which often result in her having to take time off from work. She asks whether or not she could have anything to prevent her migraines. Which one of the following drugs would you be most likely to prescribe?

A Pizotifen 500mcg nocte

B Topiramate 25mg nocte

C Amitriptiline 100mg nocte

D Propranolol 40mg BD

E Clonidine 50mcg BD

Curriculum statement 15.10—skin problems

20 16-year-old Jenny presents with her mum concerned about her acne. She has been using a topical antibacterial preparation for the last 4 weeks and, although there has been some improvement, she is still getting mild to moderate acne. Which of the following options would not form part of your management plan?

A Switch to an alternative topical antibacterial preparation as her acne should have cleared by now

B Encourage her to wash her face every night before applying the medication

C Reassure her that her current treatment may need to be used for at least 6 months to see any significant benefit

D Discuss alternative options for future management, including oral antibiotics or the combined oral contraceptive pill

E Update her repeat medication and request that she comes back to see you in 6–8 weeks

Answers:

1 B

2 A

3 B. Whilst ectopic pregnancies are relatively rare it is an essential diagnosis to exclude in all women presenting with acute abdominal pain. Appendicitis, urinary tract infection (UTI), pelvic inflammatory disease (PID), IBS and other ovarian pathologies may also present acutely in this way and it is often difficult to make a specific diagnosis, particularly if the patient presents early. Urine dipstick may well show leucocytes and/or protein in the presence of UTI, appendicitis, PID, and pregnancy and is therefore non-specific. Swabs may help to exclude PID or the presence of STIs, but these may co-exist with other possible pathologies. Abdominal and pelvic examinations may be similar in all of the above conditions and the lack of/presence of a fever isn't necessarily helpful in this case.

4 D

5 D. This lady is clearly post-menopausal having not had a period for 18 months. There is nothing in the history to suggest that she has had a hysterectomy and she will therefore require a combined hormone replacement therapy (HRT) preparation in order to protect her endometrium. Because she has not had a period for 18 months, the most appropriate preparation would be a continuous combined method as the sequential preparations will cause her to see a return in her periods.

6 E. This couple do not yet meet the criteria for NHS referral to a CRM (centre for reproductive medicine). A repeat semen analysis will often be very different to the first sample taken, particularly if the patient is willing to make lifestyle changes.

7 C. Cervical smear samples should only be taken under the recommendation of the cervical screening programme.

8 A

9 B

10 E

11 C. PSA test results should be taken in the context of a rectal examination and the patient's history. Remember, however, that performing a rectal examination is likely to stimulate the prostate and lead to a false rise in the PSA level.

12 A

13 B

14 E

15 C

16 E

17 C. Acute otitis media is the commonest cause of severe aural pain in small children. Many infections, especially those accompanying coryza,

are caused by viruses. Most uncomplicated cases resolve without antibacterial treatment and a simple analgesic, such as paracetamol, may be sufficient. In children without systemic features, a systemic antibacterial may be started after 72 hours if there is no improvement, or earlier if there is deterioration, if the patient is systemically unwell, if the patient is at high risk of serious complications (e.g. in immunosuppression, cystic fibrosis), if mastoiditis is present, or in children under 2 years of age with bilateral otitis media. Perforation of the tympanic membrane in patients with acute otitis media usually heals spontaneously without treatment; if there is no improvement, e.g. pain or discharge persists, a systemic antibacterial can be given. Topical treatment of acute otitis media is ineffective and there is no place for drops containing a local anaesthetic.[1]

18 E

19 D

20 A

REFERENCES

1 *British National Formulary*. 12.1.2 Otitis media. Available at: http://bnf.org/bnf/bnf/62/5628.htm?q=otitis%20media&t=search&ss=text&p=1 (accessed 9 February 2012).

CHAPTER 26

Top ten tips for VTS success!

Hopefully by now you should have all of the inside information required to get you through your VTS years not only successfully, but with some fun along the way. If, however, I was to give you my top ten pieces of advice then they would be as follows:

- **Make contact with your educational supervisor as early as possible:** whilst they are predominantly responsible for monitoring and evaluating your WPBA, they can also offer a significant amount of pastoral support and careers advice along the way. Creating a good rapport with your ES and proving to them your commitment and motivation for the scheme, is essential in order to make your journey from ST1 through to ST3 as hassle free as possible.
- **Be disciplined about regularly adding entries to your ePortfolio:** when it comes to ARCP panel deadline time there will undoubtedly be a flurry of activity on the ePortfolio amongst most trainees. However, this is very evident to those sat on the panel and will not create a good impression. Regularly adding good quality entries, which can be easily linked to the curriculum statements will make an appearance before the panel much less likely and will prevent last minute nagging from your educational supervisor.
- **Make the most of your hospital posts:** it can be tempting to wish your hospital posts away, counting down the days until you have completed your night shifts and can again have your social life back at the weekends. Remember that the time spent in each speciality will be invaluable to you once you're out in general practice if used correctly. Try and make

sure that by the end of each post you feel confident managing the common presentations that you will encounter in primary care and keep a note of local management guidelines that you learn.

- **Don't rush into taking your exams thinking that you can just 'get them out of the way':** you will only 'get them out of the way' if you pass and if you end up having to re-take it can be a costly business. Make sure that you are fully prepared for each exam before you commit to taking it. All failed attempts are logged on your ePortfolio and will be obvious to those making decisions about your future training if you end up in front of the ARCP panel.
- **Try to attend as much of the VTS teaching as your rota will allow:** throughout the 3 years of your training, there are bound to be some genuine circumstances which prevent you from being able to attend your regional VTS teaching. That may be because of rota constraints, illness, annual leave or exams; however, it is therefore essential that you attend these sessions when you can. Not only do they provide an opportunity to learn about some of the more complex parts of the curriculum, but they are also an invaluable way of networking with your colleagues and finding out about additional training and forthcoming jobs. Regularly attending teaching will also help you to build a relationship with the training programme directors who will then be more willing to help and support you should you run into any difficulties during your training.
- **Make sure that you are seeing plenty of patients in your surgeries and practice consulting in 10 minutes:** whist you are likely to start out in general practice with appointments lasting 20–30 minutes it is important that as the months go by you are gradually reducing the length of your appointments and increasing the number of patients that you see on a daily basis. This will not only prepare you for having to consult in 10 minutes in the CSA exam, but will also prepare you well for real life when you complete your training and start working as an independent GP.
- **Don't waste time on book work in the lead up to exams—instead use real life prompts to expose the gaps in your knowledge:** it can be tempting to request annual leave or study leave in the run up to your exams in the hope that you can fill your brain with some last minute book cramming. However, this will be of very little benefit to you compared with spending time in general practice 'learning on the job'. Gaps in your knowledge will be highlighted on a daily basis whilst seeing patients and therefore using these opportunities to form the basis for your learning will become a more natural and memorable way to study. It will also offer you the opportunity to add reflective entries to your learning log and think about future additions to your PDP.
- **Keep a dated record of any interesting patients that you see or any secondary care referrals that you make.** Learning through reflection is an important part of the training process and allows you to truly understand

whether your management plan for difficult patients could have either been different or improved. There is nothing more satisfying than getting a clinic letter back from a consultant confirming your diagnosis and explaining the subsequent management of that patient. Keeping a dated record of these encounters makes it easier for you to add them to your ePortfolio at a later time and will also make you appreciate the timescales involved when referring patients to secondary care.

- **Don't leave your OOH sessions until the last minute—the best sessions will have already been taken and you may find yourself rushing to get it completed in the middle of trying to prepare for your exams:** very few trainees enjoy their time spent in OOH, but it is an essential part of the training process. Having to spend 4–8 hours in a walk-in centre after an already busy day in surgery can seem not only daunting, but exhausting and it is therefore essential that you plan your OOH shifts well in advance. Those who leave it until the last minute may find themselves working frequent weeknight shifts or may not even be able to get enough OOH experience in order to be signed off due to demand! Try to sign up for a small amount each month, rather than leaving it until the end when you may also have exams and WPBA to be worrying about.
- **Remember that at the end of it all you will be expected to have the competence to practice as an independent GP, which includes finding yourself a job!** Whilst at the start, the end goal may seem all about passing the AKT and CSA you will soon start to realize that the VTS scheme is about so much more than that. It's about working hard to make sure that you make the most of every opportunity offered to you and providing you with the foundations upon which to build your future career as a successful GP!

And finally ... good luck, and don't forget to have fun!

Index